God's Works
In The Universe

By Dr. Hans Petermann

ISBN-13:
978-1530620050

ISBN-10:
1530620058

Nelson Publishing Solutions
392 E. Stevens Rd. G13
Palm Springs, CA. 92262

Dedication & Memorial

Dedicated to Nikola Tesla and the Aether confirming that Einstein's theories of relativity are wrong!

TABLE OF CONTENTS

What is NASA hiding? NASA = Never A Straight Answer because all they want to do is obfuscate and never give a straight answer. Also, our solar system is inhabited on Mars, Venus and Saturn.

Please note the new publisher of this book removed the old formatting and page numbers. Thank you!

Press, Kempton, Illinois 2009 Concluding dedication to the late Phil

Schneider.

Photo + copy of dead gray alien being, 1979... Photograph of ET "Diplomat" & data about it

NGC 1309 Spiral Galaxy (courtesy of NASA;)
Hubble space r-enter

Rear Admiral Richard Byrd's exploits, flying into north polar opening and over the Antarctic. Marshall B.
Gardner's concept of the north polar opening & central sun upon entering the Hollow Earth

 Hitler's ashes, UFOs and secret warfare

Lake Vostok (East) mystery + anomaly

Extraterrestrials, the Vatican + earthlings, by late
Zecharia Sitchin in interview, 2005

Concerning "Jesus Christ" and the Biblical   Patriarchs

Story of a Shaman 'CREDO MUTWA'

The mysteries of Mars, also see 'GRAVITY, MATTER +
Space Travel ' 2006, epic volume, with photos about Mars
+ data

The coming solar shift + future natural disasters   Color photo of atomic bomb being detonated e Bikini atolls, taken

August 06, 1945 and warnings for future of mankind.

Asteroid hit! Planet Jupiter once again scarred and data about Meteor Crater near Flagstaff, in Arizona
desert Off Interstate 40,

Asteroid strike on planet Jupiter and photo of Meteor Crater in Arizona desert.

Courtesy of 'Atlantis Rising' magazine @ ATLANTIS
RTSTNG.com

TOTALLY AWESOME PAINTING OF THE INNER
EARTH'S SUN REFLECTIONS AT THE

'PETERMANN'S PEAK" AREA TN GREENLAND, LARGEST ISLAND IN NORTH POLAR OPENING ENVIRONS.
WORK BY "ANGEL IN DISGUISE".

Bryce Canyon Nights " The Starry Milky Way
Rising over Sunrise Point" by Wally
Pacholka, 3—time winner of TIME—LIFE "Picture of Year"
Award wall @AstroPics.com — phone: 562—397—0591
"YOUR BODY'S MANY CRIES FOR WATERII - transcribed by this author from Trevor J.

Constable's book review of book by
Feyedoon Batman lidj, M.D. 5 pages

CHEMTRAILS, HAARP

AND GEOENGINEERING, chapter 4 pages.

PRELUDE TO "GOD'S INTERDIMENSTONAL

OMNIPRESENCE" FOR THIS IS a very extensive summary SPACE TRAVEL + UNKNOWN MYSTERIES REVEALED pertaining to very important assets and characteristics of humans residing in these very chaotic years on the surface and in various underground locations on our very active mother planet "earth" from years 2011 through the many earth changes until year 2062.

This scientist needs to emphasize the great importance of many new developments in the fields of zero—point energy and 'pulsed plasma physics, originally developed by the great scientist, Nikola Tesla and others, including T. Henry Moray of Utah and especially Philo Farnsworth, the inventor of television and plasma fusion devices.

This book is also dedicated to my late friend 'Rolf

Schaffranke', or 'RHO SIGMA' Of Georgia. His book "ETHER TECHNOLOGY" concerns itself and delves into international efforts at gravity control and discoid propulsion systems. Before the 'Quantum Field' there was Nikola Tesla and the "Ether", as mentioned in a chapter in this volume pointing out the very

important data of this field of gravity control. Rho Sigma has chapters on

'John Searle' and "Searle discs, T. Townsend

Brown and his work on anti—gravity and ether— vortex turbines. There are 3 pages in my present volume pertaining to a propulsion system using a very sophisticated 'plasma fusion' spacecraft— ether—vortex fusion chamber. Due to 'cosmic top secret' data, the development of this space technology project can't be revealed at this 'space—time'

Finally, this volume also dedicated to the late
Swiss engineer,

Paul Baumann, who developed the famous Swiss ML converter, 'Testatika machines in the early 1980s @ Methernita village in Switzerland.

Their output 3 kw/h and there also is a 12 kw/h converter operating as far as can be ascertained by this scientist, since 2002.

Please view data concerning this device on the internet @
'testatika machine', Peter Lindemann, "The Free Energy Secrets of Cold Electricity" @
http://www.teslasociety.ch/info/NTV_2011/free.pdf

The famous Swiss ML converter was and still is a self— running energy 'testatika' device, invented by the late Paul Baumann, that directly output up to 12 kilowatts per hour of electricity. These devices are used to provide electrical power to the small 'Methernita' village community in Switzerland. Since they felt mankind was not yet ready to receive the discovery of this energy, they withheld technical data from the numerous visitors and witnesses worldwide. Counter— rotating acrylic disks with metal segments like a Wimshurst machine appeared to power the devices.

The disks produced high electrostatic voltages and very colorful swirling plasma. It directed the voltage onto two sealed cylindrical chambers described as "Lay den jars." Although the many witnesses were free to examine the rotating disks; the top secrets were not revealed. . . Anyway, this device's 'secrets' can be explained by creation of cold current. Each pulse propagated through a stator system of permanent magnets, jumped across a spark gap to energize rotor permanent magnets, returned back across another gap to excite a different pair of stator— rotor magnets, and then was redirected back to the batteries to recharge them.

Magnetic repulsion between the stator and rotor magnets propelled the rotor. Here was a motor that was driven by unusual 'cold current' pulses, which displayed very spectacular efficiency at 99.99% output.

A band of glow plasma is induced in a double wall grid around the chamber's circumference, and constant pulsing activities trigger ions in motion. The resulting polarization pulses can be stepped down in voltage and rectified onto output super— capacitors by electrical engineering techniques.

My past and present laboratory experiments have already confirmed that these 'plasma—fusion— ion' surges really do indeed couple 'zero—point' energies, Therefore, this very unique Swiss ML converter is absolute proof that " free— energy " inventions can be explained from this valid hypothesis as an important key activator for manifesting anomalous plasma energies.   I would like very much to hear from the readers, as to their interest in these fields. With additional reading and new scientific data, one will soon learn the full importance of new technical data and other huge responsibilities of our present— day civilizations to bring about true peace.

If enough people are informed of the truths regarding space, our societies will then make a turn, and this planet Earth will go on to become a true space civilization in our solar system.

WHAT NASA IS HIDING INTERVIEW WITH JAMES MCCANNEY

Through a series of events and interventions that apparently were "meant" to be, on March 16, 2003 8 years ago Rick Martin of the defunct 'Spectrum' magazine spoke with astrophysicist James McCanney

•

Rick Martin: Before we get started, let me just ask what your thoughts are about the research of Dr. Dmitriev? Are you in tune with what he is saying?

James McCanney: Oh, yes. Dmitriev is an experimentalist, partly theoretical physicist, but mainly he was an observational scientist, an atmospheric scientist. He's the one who discovered the tornadoes on the Sun, and all kinds of stuff. He talked about the vacuum domain and what they were measuring in the atmosphere, and other things in the cosmos that basically they didn't know how to explain.

They were measuring them, but they didn't know what was going on. But then, when they got my papers, they said: "This explains everything we 've been seeing. So it was quite the thing. He was head of the group that I worked with back in the 1990s in Russia.

Then NASA went over there, around 2000. That's when Russia, of course, had no money. These guys were making $75 a month and were trying to publish their own work, using money out of their own pocket; it was just ludicrous. But NASA went over there, started pumping some money into them and then said: "If you teach any more of McCanney's stuff, we t re cutting you off.

"Martin: Do you have any theories or information about who specifically at NASA is behind this sort of diabolical withholding of knowledge? McCanney: Yes, it's very clear; I've known this for a long time. It is the control of space.

Martin: Can you talk about it?

McCanney: Sure. NASA is a group of scientists. That's what we always think of: these engineers who build spacecraft and that type of thing.

NASA is owned and operated by the NSA (National Security Agency)

There's a layer above NASA that controls NASA. Daniel Gold in, who came into NASA in the 1990s, came in from the CIA, and his job was to secret ise or put the cap on NASA. What he did is, he went in and the first thing he did was make everybody top, bottom, sideways who worked for NASA — sign, basically, an NSA non— disclosure agreement. The NSA is part of the overseeing government that is already in place. that's what all of the stuff going on now is about.

Martin: Are there Jesuits behind all this?

McCanney: Jesuits? (Laughs) The Vatican has a big stake in the world— wide government, and its part of it but not the whole show. It's very much a worldwide situation, where you literally have hundreds of families who are associated with this. They are very wealthy; they're in every country of the world; they control the politics and the money and the banking. So, it takes a very large web of these people. Martin: I was going to mention the Nazis; that's where I was going with my original question.

McCanney: Yes, many of the Bush Administration are either direct descendants of Nazis or of those who helped finance the Nazis, They, of course, realized that space is the last frontier in resources. The control of space is essential to everything that they're doing. It's the last frontier. "PLANET X" AND ITS COMPANION COMETS

Martin: What is the fear of NASA concerning "planet X"? Is it related to Sumeria and the Annunaki? Or is it something else?

McCanney: I wouldn't say that, but the knowledge that there is this "Big Thing" that comes in on a regular basis is old. That's part of the very high levels of secrecy in a lot of these groups, like the Vatican. I mean the first thing when Hale—Bopp showed up, the Vatican built a world—class observatory in Arizona and staffed it with astronomers. Gee, wonder why? Then they have a second one, but what is interesting was after Hale—Bopp left — because they thought that Hale— Bopp was the Big One. Now, let's backtrack to 1991. Hale—Bopp was officially discovered in 1995, by Alan Hale at New Mexico, and then by Bopp, the Japanese guy. They both hit on the same night so they both got credit for the name of the comet.

Prior to that, it's very clear to me, and if you're looking at my Harrington notes, that one of the things which Robert Harrington was looking at was Hale—Bopp. The nucleus of Hale—Bopp was extremely large. The reason NASA pulled the feed down immediately, once they realized that some lackey had stuck it up on the Internet, was because any astronomer or person like myself would know that with that small amount of data you could determine the mass of the central nucleuse It t s a little equation you use. They use it all the time to determine the mass of central stars when they see a wobble in a star; then they can determine the radius of the thing orbiting it. You need the period and the radius of whatever is orbiting the larger object, and with those 2 parameters you can calculate the mass of the central object.

Just a little equation in celestial mechanics.

So, with that small piece of evidence on the web, anybody could calculate the mass of Hale—Bopp, showing that it's planetary in size and coming in. Now, the other factor. In 1991, what Harrington saw was 2 things: he saw Hale— Bopp, and he saw something much bigger beyond Hale— Bopp. That's planet X. That's my understanding at this point. In 1991, Hale— Bopp was on a near direct collision course with earth, with a couple of quick photographs they could chart the orbit, and it was on a near— collision course for Earth.

Martin: No wonder there was such a scramble.

McCanney: It was a huge scramble. When it was discovered, I called up Goddard Space Center I knew the secretary there and I said:
"What's going on? I heard there's this comet. "You could hear the screaming in the background.   And she said: Oh my God, this comet is huge!" But I thought she meant in the sense of being a news story. No, it was huge in the sense that it was a planet— sized object. They had been tracking it.

You see, this is where the division comes in, because it wasn't until then that even a lot of the scientists at Goddard found out about it. But they had been tracking this since 1991, possibly earlier. Harring— ton discovered it, and you see it in the 1991 memo that he knew exactly where to go and look at it.

A long time ago what happened is...it was one of the companions of Nibiru that did the damage to Earth.
Martin: A companion?

McCanney: A companion. It was the one that became Venus. Velikovsky was very right that Venus was a huge comet that worked its way through the solar system, and it took about a 600—year period from the time it was captured by Jupiter to the time it encountered Earth, and then worked its way in to become the planet that we know today,

So originally, what happened was that Hale— Bopp was here about 4,200 years ago, and Venus was captured by Jupiter about 4,200 years ago. They were, literally, smaller companions to Nibiru, that's why they didn't want anybody to know about the companion, because they knew it was on a collision course with Earth, and they knew it was the companion to the bigger one that caused the problem. But they didn't realise that Hale— Bopp was, literally, one of the companions itself. Now, when the destroyer, the Big Guy, Nibiru, comes in, it has an entire entourage of these things.

Martin: I guess comet NEAT would be one of those?

McCanney: And that's the thing. When we got barraged a few weeks ago by all these comets, and they never announced Comet NEAT, C— 2002/VI.

Clearly, all of this stuff is coming from the southern hemisphere. Then, of course, Harrington knew very well where that was, for the reasons that I gave; they were pulling down on the planets Uranus and Neptune. It's interesting to note that when the story of
Harrington came out, the government tried to make a statement through some of these astronomers that are on the radio, the disinformation guys, who came up with a story: "Oh well, we have corrected the masses of those planets due to new information, so that has taken care of that problem."

Well, no, that doesn't correct anything when you see these planets being "pulled down". That would only correct things in the plane of the planets. This object was big enough, back in 1991, that it was pulling Uranus and Neptune down out of their orbit. That's how big this thing is!

So, you see the concern over the companion. Because they all know, and the Vatican knows, that it was the companion that did the damage the last time. The only problem is; the companion became the planet Venus. What they don't understand is that it's a very difficult thing to produce the orbits for these, and NASA is learning that the hard way. They couldn't keep track of Hale—Bopp; it changed on a daily basis. That was one thing he did in the Millennium Group: track the daily changing of its orbit on the government ephemeris pages. Martin: Was comet NEAT a surprise? Did that come out of nowhere, or did they expect that?

McCanney: No. Comet NEAT is another very large nucleus; planetary in size probably the size of our Moon, at least; probably larger. NASA knew it was coming. They probably saw it coming in years ago, as part of this entourage of things coming in which I think of as things that are coming in as part of the planet X entourage. They didn't want anybody to know about it, for the simple reason they knew it was going to come in right around the Sun and it was big. They probably never expected it to become as bright as it did. But it was literally visible in the daytime sky, right next to the Sun, as it passed over about a 12—hour period when it was coming in.

Martin: The obvious question is: how many more of these companions can we look forward to?

McCanney: We don't know. Of course though, NASA would have very good knowledge of this.

The other important thing that I wanted to say earlier about Hale— Bopp is that in the six—year period from 1991 to 1996, where it actually hit perihelion with the Sun, it lost 3 months' time in arriving due to the tail—drag of the huge comet. That's why we didn't have the direct collision with it. And when I say direct collision, I don't mean hitting Earth; I mean we would have been within about 1 million miles. By anybody's standard, it would have been total devastation of this planet. The flooding would have been enormous. It was only due to the fact that this thing slowed down that we missed 199

Martin: I guess another question would be: where is NEAT going?

McCanney: NEAT headed back out. That is object number of my South American Harrington expedite o to chart the new orbit for NEAT, because it clearly of its energy as it came around the Sun, as it picked up a lot of tail material.

So, it's not going to come and hit Earth. That's what NASA always says: 'Oh, these people think it's going to hit Earth." No, no; nobody said anything about it hitting Earth. They try to make fun of people and, in fact, they actually have people who set up those stories on the Internet so they can go make fun of them. It's part of their disinformation campaign.

Martin: When a comet the size of NEAT, or a planet the size of NEAT swings by our Sun, how does "action at a distance" come into play? McCanney: The flare that came off the Sun, that you see in many of the photos, came and hit the back side of the comet tail.

Martin: The five—million-mile flare?

McCanney: Yes. Now, if that were to have come at Earth, it would have knocked us to our knees. But it went off in a totally obscure direction. Let's look at something else. What you didn't see there, but I could see it coming off of NEAT If you look very closely, you '11 see a pin—thin streak coming directly away from the Sun and out away from the nucleus, out the right of the screen. That's connecting with planet Mercury. Mercury was in a direct alignment with NEAT as it came across the ecliptic, the plane of the planets. That line, you can actually see on solar photographs, is connecting to Mercury.

So, now, let's put Earth over there. What if Earth had been over 90 around, and we were not broadside to it? Then, we could have very easily been in a position to take that flare, for example, or take an electrical discharge directly from NEAT e That is what the ancients talked about with the comets, the lightning bolts flying across the heavens; they saw these things   Zeus throwing lightning bolts to Mars. They saw this stuff.

Martin: It was literal; it wasn't metaphorical?

McCanney: No, it was not metaphorical. When Venus came around Mars, floating around the planet. Things just didn't make sense. There was a proton wind. The thing that caught my attention the most was that there was a proton wind coming off of both Jupiter and Saturn, and that's a satellite property that we only see from the Sun. It's interesting that they only saw protons; they didn't see an electron wind that would neutralize that.

So, clearly, outer space was not what they were expecting. They were expecting Jupiter to be an ice— cold frozen ball of hydrogen, very sterile   which it is not. It is tremendously dynamic; has a huge magnetic field. Literally, as they went out in front to Venus, as they went out to all of the other planets, they found them very different than what they thought they would be.

At any rate, T happened to be at Cornell at this time, and by then I had already completed much of my theoretical work on inclusion of electromagnetic fields and celestial mechanics. I understood how these worked. When I saw this data coming in, I recognized it and that, of course, is something that none of these Chapman physics guys had any clue about. They were still trying to imagine that these were gravitational effects that they were seeing.

At that time, I also studied comets as part of what I was doing. I realized that comets could not possibly be these dirty snowball things. There was a lot of data indicating that comets were inter— acting electrically with the Sun, and they were noticed to have electrical discharges around them. At the time, I didn't know what caused the electric fields, but I knew they had to be caused by the Sun. I knew that the comets were interacting and that the nuclei of the comets were becoming negatively charged.

Then, it finally dawned on me at that time, 1979— 1980, that this had to be produced by a differential flow in the solar winds. In other words, there were more protons in the solar winds than electrons. That gave me a whole new model for fusion. That's when I realized that the fusion had to be up in the solar atmosphere and not down in the core. That's when I realized that the corona of electrons around the Sun was really a super— atom space, and that the Sun itself was positively charged down below that, and up above that the corona of electrons was actually making the Sun look negatively charged to the outside. This whole complex phenomenon of how the solar winds would open up holes in the corona and come blasting out was caused by electrostatic acceleration of the protons as they moved out through the corona. And that's exactly what we 're viewing. And this whole time, even up until today, NASA insists that the energy from the Sun is coming from the core. Totally incorrect.

When I was at Cornell, I met Hans Bethe, Nobel — Prize winning physicist who created the model for the Sun that we now use. And, of course, he was a friend of Albert Einstein, and they both won Nobel Prizes. Hans Bethe won the prize for the chemistry and the under— standing of the nuclear fusion model that we now use today, that the chain reactions would build up the bigger atoms and cause the heat. He made the initial calculations that the heat of the Sun, and those kinds of things, would actually match reality.

I sat and talked to him about this. I talked to him about the fact that the solar system had to be electrically active and that comets were not dirty snowballs. And he looked at me and he knew, and Einstein knew, that. .one of the last things Einstein did was very actively pursue Velikovsky's work, because he knew that General Relativity was missing something very big, and that was the electro— magnetic field. You could not have gravity affecting light without also having the electromagnetic field around stars affecting light as well. He knew that those factors were missing from General Relativity, and that's what he was working on when he died. Hans Bethe told me that's what Einstein was working on; he was trying to figure out that problem.

I asked him: "I'm having trouble publishing. They're taking away my ability to publish. Do you have any suggestions for this?" And he said: 'Try the German publications. " And I did. My work eventually began to be published in The Netherlands.

Martin: That's interesting. So you had to go offshore.

McCanney: Yes, and there were two journals that were published in the
Netherlands: Astrophysics & Space Science and another one called The   Moon and the Planets. This
contact was due to what Hans Bethe told me. Martin:
Sounds like he gave you good advice.

McCanney: Yes. But when this stuff started hitting the streets, the people at Cornell freaked out.

Martin: Why is that?

McCanney: Because I was using Cornell's name, and I was using non— Chapman physics with Cornell's name on it, this was not what they wanted to see. And, of course, when they got a hold of all of my papers and ran them through the Space
Science Department there, they realized that what I was doing was corroborating
Velikovsky t s story. Carl Sagan was Professor
Emeritus of the Donald Duncan Chair of

Astronomy, a very exclusive seat of astronomy at   Cornell. He was the one who, basically, did in Velikovsky. That's why Sagan was famous. Not many people understand that he led the charge against Velikovsky, who was selling millions of books all over the world. Sagan led the charge that Velikovsky was a geologist and planetary scientist and astronomer, and on and on, to prove that Velikovsky's thesis could not possibly be true.

And that's why Sagan eventually got the Cosmos series, because he was the spokesperson for the astronomy community that buried Velikovsky. Not more than 2 years later, T show up at Cornell, using their own data to prove Velikovsky correct. EFFECTS OF "PLANET X" FLYPAST

Martin: Let's talk about "Planet X" some more. I know you don't like to talk about time frames, but do you have any sense of it at all?

Are we a year out? A hundred years out?

McCanney: That I don't know, and that's what I want to find out with the Harrington Expedition.

Martin: So, you don't have a sense of that, at this point?

McCanney: I do, privately. I'm always afraid to speak of dates because people try to hold you to that.

Martin: Could you talk in general terms?

McCanney: Okay. If history plays out correctly, let me say this. Hale—Bopp, NEAT and the other 5 comets that we saw in February... we saw 5 comets come in very close to the Sun: one was Kudo— Fujikawa,

 one was NEAT VI, the other one was no name it just came tunnel ling into the side of the Sun at about 100 million miles an hour and then there was another small comet nucleus that tunnel led up into the Sun, on the lower—left side of the picture as you look at it, as NEAT passed on below. It is believed that this was an object that was thrown off of NEAT and tunnel led down into a solar flare — and that's where those big balloon— shaped, long plasma tubes came out of the Sun, in reaction to that.

Okay, you question was about Planet X. The Hopi

Indians believed that Hale—Bopp (1995) was the Blue Kachina, which was the precursor by about 10 years of Planet X.
 And 10 years, of course, is a very relative term there. The point is, if Hale—Bopp had been a companion of the Big Guy 4,200 years ago — and that's what the cycle is: not 3,600 years, but 4,200 years for Nibiru — then it's due here within the next decade. And the other thing is, people are concentrating on this Planet X or Nibiru object. The thing I point out is, I study the extra—solar system objects. NEAT, for example, didn't match anything that

we've seen before. It was a brand new comet. So, whether it was related to Nibiru coming in, or not, is impossible to say.

Martin: So, really, it's an unknown entity.

McCanney: Right. The point is, there are hundreds, if not thousands or hundreds of thousands, of these big objects floating around out there. And that is something that NASA refuses to acknowledge.

Martin: The so—called "panic theory"?

McCanney: Yeah, and then that logical deduction. "If that came all of a sudden out of the blue, could another one come out of the blue, at any time, and come by Earth and affect it?" And of course, that's what I t m saying. That's the "action at a distance" thing. We don't have to be hit by these things. NASA keeps pounding on this: "If we're hit by one of these things." No! It has nothing to do with necessarily being hit actually.

If that flare had hit us, had come out and gone out behind NEAT, you would have known it. In 5 years' time, there'd be a lot of people dying of cancer. because it would have blasted the magnetic field, would have torn away our magnetic shield, and whoever was facing the Sun at that time would have been toasted. It's not well known, but back in the late 1990s this I got from Dmitriev, in fact there was a solar flare that hit Earth, and we (the USA) were on the night— time side at the time when the magnetic field went to zero. Russia was pointed toward the Sun, and they' re actually tracking cancer rates now in Russia, based on that flare. Martin: That's amazing.

This will get into an area that you might not feel comfortable answering, but my wife is curious to know what effects on people these vibrational changes will have over the next few years.

McCanney: I would say twofold. There is very much a polarisation going on, right now, around the world. You have the people who are raising their consciousness of understanding of where we fit in the Universe, that know we have to come together in peace and stop using the resources on this Earth in a totally careless manner. We have to provide for our future generations. Those people are going to be elevated, way up. And they're going to be communicating on an almost mental, spiritual level around the world, understanding that we can't continue to have petty Earth—wars and put all our resources into this. The other half of the polarisation in this global organization that is controlling the whole world and keeping it, basically, in slavery mode. These organizations and 'elite families' are going to become far worse in doing what they're doing. They are totally dedicated to doing nothing but war and destruction and killing.

Martin: Let's talk about the shifting magnetic poles of planet Earth, and how these changes are affecting our magnetic poles?

McCanney: First of all, the magnetic field of the Earth is very much misunderstood. Most of it is caused by currents that flow around the Earth. It's not caused by some kind of magnet in our core. The magnetic that does come from our core, the permanent component of that magnetic field, is very loosely bound in iron and nickel deposits. It's not like a little iron magnet that you would put in your pocket as a kid. Most of our magnetic field is in the form of electrical currents flowing around the planet in the solar wind, and in the Van Allen belt, and in other forms that's our magnetic field, that's why, when a very highly charged electromagnetic comet comes by, it can very much affect us.

This is a good time to mention that the Russians did a study a number of years ago on foetuses. What they realized was these unborn foetuses were tuning in to the electromagnetic rhythm of the cosmos. The Russians were very aware of the electromagnetic part of our environment, whereas in the West, they were still saying There is no such thing"

Then they did statistical studies on the planetary alignments, and things like that, relative to astrology. And, basically, they became convinced that there was a very definite association with people, their lives and the way they acted, and the planetary positions. They did this with not just humans but plant life, animal life, and on and on.

They realized that there was something to this, but they didn't understand what it was.

But when they realized that all of the planets are discharging the solar capacity that's built up around the Sun, due to excess currents of protons in the solar winds, they then could see that as these planets came into alignments this increased the flow of currents along those paths. And when you had many planets line up, it increased the currents 100—fold, not just twofold.

And so, as the Moon, for example, goes through a New
Moon phase and passes away, for a short while in the New Moon phase it's blocking the solar wind. But as it moves out of the way, that solar wind comes pounding in and breaks our magnetic field down, causing tremendous pressure on the atmosphere.

The New Moon phase and the Full Moon phase are times when the Earth is being, basically, crushed under a lot of electromagnetic pressure, which is just one way of talking about it. So, all of these effects that you 're seeing are very real. When a big comet comes into the solar system, breaks down the solar electric field and starts driving plasma up there; it's doing strange things; we don't know what causes all of this, but space is not electrically neutral it's very much active electromagnetically. The United States, of course, is where the money is — so Chapman, the American, won out.

There is a very simple physics problem that is taught to every graduate student in space science, astrophysics and physics. That is, if you take a charge and put it in outer space, then very, very rapidly — and you can calculate how rapid this is charge will come and surround it and shield it, and will not allow it to be seen, electrically, in any other part of the Universe. It's a shielding property, and if you have a magnetic field out there for some reason around an object, the same thing will happen; you get a plasma effect. And that is, for example, one reason why our gravity is known to be a force that is totally independent of electromagnetism, because these electromagnetic forces are so shielded that gravity "sees through them", let's say.

Be that as it may, Chapman kind of won this theoretical battle. And so, for decades you had the Chapman conferences, and Chapman physics was taught in all the textbooks, and all of these guys grew up thinking that space was electrically neutral — because of that little problem you could do as a graduate student. And I've done that. But what I realized, and apparently none of these other people realized, was that the data, as it started coming back from the space probes, didn't support that at all. There was a tremendous amount of electro— magnetic activity out there. That was in 1979 when I was a young instructor at

Cornell University and had access to all of this data coming in from Voyager spacecraft, pioneer and Voyager, as they went by Jupiter and Saturn. That was before NASA realized that they had to keep the data away from people like me who would interpret it differently than what they would like to see.

The data was saying something totally different. Very bizarre electro— magnetic things were being observed: spokes in Saturn's rings, tremendous electrical discharges, current rings of millions of solar flares lashed out with an electrical discharge and the auroras in the atmosphere of Mars lit up; it looked like a snake grabbing Mars. It literally sucked the oceans and atmosphere off of Mars as it passed by. And they saw this. They knew that Mars, prior to that, was a water planet, was a blue planet, just like Earth. Mars has a Very thin atmosphere, Venus has a massive atmosphere, thousands of times denser than Earth's atmosphere. But percentage—wise, the chemical composition of the atmosphere of Venus and Mars are exactly the same which means they were formed in the same boiling pot there, as they passed by each other. Martin: I wanted to ask you about Velikovskye you were very similar to him in that he was given a hard time and ridiculed, and it turns out he was right.

McCanney: There's no question any more that Velikovsky was right. And, I think the biggest thing that I want to say about Velikovsky is that he was not studying astronomy. He was studying calendars!

COMETS AND PLASMA PHYSICS

Martin: Let's talk about your concept of comets and plasma, what is plasma?

McCanney: Plasma is like a fourth state of matter. In a vacuum environment where you have strictly gases and high energy, for example, a lot of light coming out of the Sun that splits the atoms into free electrons, ions, neutral atoms and other forms of energy like stored magnetism, stored electric fields that's a plasma. And the inter— action of all of these things is what you call plasma physics; that's the study of it. That's literally, in a nutshell, what plasma is. Martin: Let's talk about comets. They 're not dirty snowballs. What are comets?

McCanney: Let me start by saying this... For a long time, there was somewhat of a feud in the astrophysics community between a guy named Chapman and Hans Alfven, from the Swedish Institute. Chapman said that space is electrically neutral. Alfven said no; we can see this energy up in the Sun, as we're seeing, then all of a sudden you get what I'm talking about with this polarisation. The people who are raising themselves to a higher spiritual consciousness will raise themselves much farther, and the people who are intent on war will go out and beat the war drum much more. And that's what we 're seeing right now.

Martin: What do you think individuals will experience? Will people experience much more disease, be more out of balance any theories?

McCanney: Let's talk about the jet stream and weather. Everything will become more erratic. We're already seeing this. You '11 see temperature swings that are much larger. And you're going to see the same thing in people. People who are not really in control of themselves you 're going to see become erratic; people who are off balance a little bit become off balance a lot; people who are very balanced become far more balanced. So, this is part of the polarisation. There's not going to be anybody in the middle any more.

ETHICS OF THE NEW SPACE RACE

Martin: Let's get back to the "zero point", a politically incorrect subject.

McCanney: Here is what I think is going on, and I learned this when I first met the Russian people. They would talk in terms like.

Dmitriev talks about the "vacuum domain", and that, to us, means outer space.

When I met with Valery Uvarov from the National Security Academy of Russia, I told him: "What they don't understand in this country (USA) is, there's a higher level concept. When you get up to this kind of hardware let's call it hardware, because that's what we would call it in this country — you can reverse—engineer and you can have all of the knowledge on how this stuff works, but it will never work properly if you have evil intentions. When T told Valery this, his eyes got bigger than saucers. He said: "You understand this? You 're the first person in the West who understands this concept! "

Martin: Those things will not be allowed in space with evil intent? McCanney: No, absolutely. And the people in "black ops" here have gone to the extent of going to the East and actually getting higher level consciousness humans to come over here and stand next to their space— craft to try to induce that consciousness into what they're doing. And this is something that is a real problem in the West, because Russia already knows this. They have broken the ties with that kind of philosophy and are moving on. Valery told me; he said they are accelerating; they are being contacted. He said very plainly to me: "Your country and the people in it will not be contacted until you break that down." Because this is like a lead weight on the planet, this country. 'Then you'll start to progress." It was just so obvious. And those Russian people are making great strides. It's not because they have a hundred— billion—dollar budget. You don't need money. These devices are not complicated. It's very clear that the ancients had space travel, and they had the understanding of how to shield them— selves electromagnetically as they moved throughout space, and to move electromagnetically.

Martin: Are there some closing comments you'd like to make?

McCanney: I would say, number one, that the rest of the world is advancing far beyond the United States in consciousness and in progress as a human species.

The other thing I would say is that as a country, as a civilian population, we have to grab hold of this country and turn it around be— cause, literally, the whole rest of the world depends on it. We are at a stage right now that is equivalent to 1939, pre—World War 11 Hitler's Germany.

They did not turn that country around and if we don't turn this country around, we 're going to be in a far bigger world problem than World War 11 ever was thought of being."

This interview with James McCanney was extracted and edited from the May. 2003 issue of The SPECTRUM news magazine, out of print now!

EVOLUTION AND MAN'S TRUE ORIGIN

Darwinism only dealt with evolution of material organic forms, where— as the complete picture is one in which we must consider both physical evolution and spiritual transformations and integration. Darwin's theory is even less than half correct. Man did not evolve from lower species nor did lower distinct species evolve from one to other higher forms in the broad sense of the theory. One should think more in terms of each specific species as evolving "vertically" within its own framework of evolution and not "horizontally" from one type of creature to another.

Now we have a very different situation with consciousness or spiritual integration. But not only do we have to consider both the physical evolution and the spiritual integration but also the interaction between the two.

COLLECTIVE CONSCIOÜSNESS

Spiritual "evolution" firstly involved seemingly a devolution in the sense that consciousness descended into material forms. It started with collective consciousness. This was given the purpose of separating into many souls, each then to develop independently learning by interaction on lower levels.

Thus we have 1) the physical evolution involving gradual integration of cellular organisms from atoms and single cells, and 2 (a) quantitative spiritual segregation; division of collective consciousness into separate parts, with the possibility of devolution spiritually, also 2 (b) qualitative spiritual integration in which independent souls re— establish contact with their origin.

In addition, we have to consider interaction of organic matter, (1) above, and spirit, 2 (a) and 2 (b) above. One can see now why the subject has not been made clear. The quantitative separation of spirit and its qualitative integration are not separated from the physical changes.

Firstly, collective consciousness can be thought of as a massive volume of spirit which was given some initial qualitative independent characteristics by the Infinite, different from other species of consciousness. We can now think of this large fragment of consciousness as dividing into approximately 2 halves: one forms the human life— stream to associate with the material worlds, and the other, the life— stream of beings to be specially formed and of limited evolution, for continuous operation in the higher worlds called angels with the purpose of providing the continuous spiritual counterpart to humans, for guidance in the more negative existence of Earth lives, and eventually and ultimately reunion in the New Age.

This is not so dissimilar to the action of man focusing his consciousness (compare human life—stream) into an area, almost hypnotically, but leaving part of his attention (the angels) observing himself and thus able to pull him out of a trapped hypnotic condition (dramatization of materialism).

According to Hilarion's material this collective consciousness of the human life— stream was originally associated with lower integrated physical forms such as the air kingdom, mountain ranges, plants, and apparently even the dinosaurs, etc., all involving valid experience for this consciousness.

In the development of physical organisms, although they have their own associated consciousness they have no significant identity or individuality.

These lower consciousness' or psychic anatomies evolve and link up between lives with expanding regions of consciousness as they reincarnate into higher physical forms. Thus this type of consciousness is building up and resonating with higher levels more and more, compared with man's spirit which, as we shall see, comes down scale in its association with the physical levels. Primitive man comes into this lower animal category. Man, today, did not evolve from Stone Age man.

Now initially this human collective state was distributed into super— human type bodies of a more astral type (which were provided). The spiritual level was still only collective though. These bodies were of lighter structure, as "astral" implies, than present—day more dense bodies, and were also androgynous or hermaphroditic.

That is, they possessed a high degree of unity — above 3 dimensional — in which positive and negative, or the male/female genders were not yet expressed or separated out. Note that it is not strictly correct to say that the androgynous unity or body contained both female and male properties, as this is like defining, say, a solid sphere as made up of 2 hemispherical halves or 2 extremely complex shaped halves. It would be a non— sensical explanation the male and female are merely inherent as a particular possibility.

THE TEMPTATION

The next major event was the Temptation. Man, with his secure androgynous body, had not sufficient weaknesses or insecurity to respond appropriately to the Temptation plan or program. Thus at this point the body was divided into male and female bodies. Remember that the androgynous body would be of higher frequency and not visible to us in our present, and it did not entail anything too miraculous or preposterous to crystallize 2, more— physical forms.

 Thus in addition to presenting a body form of greater vulnerability, the existence of the sexes encouraged greater division for conscious— ness (in support of the original purpose — that of creation of individual souls). Note that this process of division apparently occurred over a period of hundreds of thousands of years as more of the race volunteered for division (Hilarion's material).

ADAM AND EVE

These were now the Adams and Eves (there were also
other series according to
The Urantia Book) and they lived in pairs in perfect harmony. Note that the Bible's metaphorical statement of Eve 's creation from Adam's rib implies a nonsymmetrical split!

The subtle procedures introduced by the Temptation are evident as we observe that all factors are constructively involved with the Creator's plan for mankind every negative facet anticipated and given a constructive path. This extraordinary intertwining of negatives and positives in which both deliberation and spontaneity are permitted in which action and   reaction or error are allowed, gives a hint to man of the workings of the Infinite, in which the dichotomies of freewill and preset organization of events or parts, integrates greater Being— ness within the One.

In the Hilarion material it is explained that it took another race of superior beings (also mentioned in the Bible), along with a number of powerful overshadowing spirits to provide the conditions of the Temp— tat ion. Lucifer, a brilliant and beautiful celestial being volunteered to provide the negative influence which was to encourage interbreeding amongst the different races.

Note the incomprehensible aspects to man's present intellect, of an angelic being, Lucifer (a positive), volunteering to attempt disruption of the Adams and Eves' harmony (negative action). In his success (a negative), since the other race had independent souls, off— spring from the interunions also had independent souls (a positive) and a further step of the Creator's goal was achieved.

According to The Urantia Book the Temptation was quite different but it is probably referring to a completely different period, one much more recent and of the order of 30, 40,000 years ago. However, it does refer to their description of Adam and Eve as being of the third physical series. In this context Eve was swayed by Caligastia, the Planetary Prince of Urantia (Earth), whose concepts clashed with the Divine plan for planet Earth. In turn, and to avoid losing Eve, Adam compromised his integrity and went along with the plot. Satan was Lucifer's first assistant who was in league with Caligastia but the latter is referred to as the "devil" and a deposed Planetary Prince of Urantia. There is no reference to the more popularly known aspects of the Temptation.

To continue with the Urantia material, Adam and Eve were created over eight—feet tall and divided from their original androgynous unity specially for their Earth mission. They were given the ruler ship in a strife— torn world. The task was particularly arduous and discouraging as a result of the effects of the earlier Lucifer rebellion which is described in detail in the Urantia Book, and relates to Lucifer's criticism of the entire plan of universe administration, although he initially retained apparent loyalty to the Supreme Rulers. He questioned the validity of the Universal Father and maintained that physical gravity and space— energy were inherent in the universe.
He did not recognize that personality was a gift from God.

Lucifer claimed that far too much time was spent on training ascending mortals, and that the destiny was pure fiction. He advocated self— assertion and the liberty of individual self— determination. This over— emphasis of self was seemingly the seed of the rebellion.

LUCIFER

The Urantia Book, similarly to other sources, describes Lucifer as a magnificent being, a brilliant personality and a perfect "created Son" of the local universe. The disastrous results of his instigation of the earlier rebellion are dealt with, expressing fully the seriousness and subsequent penalty, but with no hint that such a plan was within the constructive foresight of the Infinite, or accommodated by It, but that, in fact, the origin of Lucifer's turning away from the Light was unknown and one for speculation!

We mentioned earlier that Lucifer volunteered to play the role of mischief maker. This level of creativity and decision is very high on the hierarchical scale of consciousness and unity, and Lucifer's offer to play the "devil" would be analogous to writing a play then taking part in it on a lower level in which the lower level is a true reality — just as do human souls for the lives of the lower selves.

The crimes enacted in the "plays" must be handled as actual and are subject to strict laws at this level   all consequences and karmic dispensations must be experienced. Since existence is subject to laws within any dimension but there are different laws for different planes or dimensions, the manner by which Lucifer achieved this double role was to create a "reflection" of himself and setting this counterpart to do the evil work. Thus the original Lucifer was acting constructively on a very high level (somewhat similar to the higher constructive (positive) action of a catastrophe caused by the Creator to benefit man in the long run). He produced a negative polarity of himself which when eventually played out, or resolved out of the negativity, would unite with himself and would make him greater, but the karma would have to be administered and experienced.

According to the Urantia Book, Lucifer, which we can assume means his negative counterpart, is still on one of the seven detention satellites orbiting one of the seven planets orbiting the headquarters sphere, Jerusalem all architectural bodies, made to order...

It may be apt here to mention that Earth could be label led a prison planet — which might, however, be taken as an insult by many of its inhabitants. Purging of the galaxy of the many sources of the Dark forces have resulted in deliberate and enforced incarcerations on Earth of many fallen angels, particularly some of those from Maldek. Maldek was a planet in our system originally prepared for fallen angels to work out their karma. However, they failed, became more negative, and the planet was destroyed (the remaining massive negative thought forms were attached to Mars to avoid over— contamination of Earth, and the planet debris forms the asteroid belt, one of the un— explained mysteries of our solar system, between Mars and Jupiter).

As a result of all the negativity to which the human life—stream has been subjected in addition to that justified by man's own acts such as the above, and the enforced negativity applied to encourage formation of selves or independent souls, man, at opportune periods, is allowed relief from some of the karma he has acquired; for example, by means of the crucifixion, and certain violent solar activity as the Sun absorbs some of planet Earth's karma.

INTRODUCTION OF THE DEATH CYCLE AND
REINCARNATION

To continue with man's evolution, the progeny from the cross—breeding had individual souls (since one of the parents, an advanced being, had an individual soul), but many unfortunate mutants were created.

In addition to this inevitable occurrence of mutation, the first race of men (the advanced beings) taught certain mind powers which were eventually misused and which were more lately referred to as dark practices or black magic.

Following the first major event, the Temptation, came the Great Flood, which constructively destroyed the mutants (except for one species) that were producing further genetic havoc with their offspring and also implanted a profound awareness within man, by subconscious symbolism, of his responsibility in the preceding evolutionary events.

The mutants, of subhuman nature which survived, were allowed to do so to provide a suitable but inferior vehicle for souls with corresponding psychic energy malformations to reincarnate into, to work out their deficiencies. These humanoids gladly served the superior humans and the galactic colonies inhabiting Earth at that time.

Remains of these mutants have been found by anthropologists and designated Neanderthal man as a result of these sub-humans' occasional neglect to cremate their brethren, which was the standard procedure at that time. As we come forward through the epochs of history, the rise and fall of many civilizations, Lemur i a, Atlantis, Anti la, and many other less significant civilizations to man's present evolutionary status we see a repeat of technological advancement and spiritual retardation with the inevitable climactic changes prophesied by the Tribulation and then the Transformation.

The Tribulation is a 7—year period given in Revelations of the Bible and corroborated by extraterrestrials and the spiritual hierarchy, in which unprecedented changes occur on the planet both at the material level and spiritually, with the purpose of ending the negative cycle of man's existence and beginning a new era referred to as the Golden Age.

The Transformation is the 1000—year period following the Tribulation in which a transcendence of Earth and consciousness will occur.

PLANETARY VIBRATIONS

The planet has a relatively high—frequency fundamental vibration which, apart from the negative distortions, is linked to all its sub— harmonics in atoms and groups of atoms, and also higher cosmic frequencies such as those of the Sun, Galaxy and universe.

Up to this point man, male and female, was immortal. However, it was opportune for several reasons to introduce death or a natural dying cycle and reincarnation. Constructive reasons were that man was acquiring an overload of negativity; from the Temptation incidents, and the subsequent selfishness and control over others through the dark practices, and that the between lives' period after death could be used to advantage to keep in check this negativity.

This was achieved by both 1) the death experience of a partner and the consequent extreme loss, bringing cancellation of some karma, and 2) the increase in spiritual awareness between lives.

It was imperative, however, that the individuation of souls was sufficiently established so that it was retained after death and that the collective state did not dominate. This extra reinforcement of self was achieved by taking a part of the negative souls or fallen beings who had led man astray and add it to the souls of man such negativity and out—of—phase energies ensuring separateness.

THREE METHODS OF REBIRTH

With the advent of the birth—life—death cycle arose three different methods of rebirth into physical vehicles, depending on the physical densities of matter and living bodies. The Hilarion material explains that the early method, before further deterioration of man's consciousness had set in, reduced frequencies, increased solidity, etc., was to precipitate a body from the ethers by concentration at sacred locations, which when this procedure was no longer possible became temples of worship.

As man deteriorated, and matter became more solid and objectified, it was necessary for several souls, in assisting an individual, to unite in projecting energies for the formation of a vehicle for that individual. Finally, as this method became no longer possible, birth was achieved as at present by fetal development by means of the establishment of a genetic code in the womb.

THE GREAT FLOOD, NEANDERTHAL MAN, AND
CONTINUING CIVILIZATIONS

As the consciousness of man expands, his vibrations of mind and body rise in frequency. At the present time this is occurring with an in— creasing number of people on the planet and consequently this will influence the vibration of the planet towards greater frequencies. Any sudden changes though, even of a positive nature such as the effect of the Christ consciousness through Jesus could upset the balance of the energies of the planet. At the time of Jesus, the huge influx of negative forces (which were allowed by the Creator or spiritual hierarchy) produced a huge negative factor which balanced out the positive one from Christ.

At the present time, the balancing element is being provided by the fact that the negative elements on Earth are increasing (just as the positive ones are increasing). Bad music and art are becoming worse, satanic cults are expanding and drugs and disharmony are becoming more widespread in these lower ranges of consciousness' vibrations. During the thousand— year period planet Earth will gradually fade from the 3D space—time matrix as its lower harmonics withdraw up the frequency scale and hierarchy of levels of consciousness.

ADVANCED LIFE ON MARS AND VENUS

Cosmic visionary Ruth Norman of the Unarius
Foundation informs us in her book 'Mars Underground Cities Discovered' that Mars was the victim of a huge disturbance some 160, 000 years ago, described as a Nova but involving the entry into our solar

system of a body passing close to Mars, colliding with Earth, and destroying the continent Lemuria. We shall return to this but in the meantime let us take a look at contemporary science's information on Mars.

Scientists continue their age—long quest for the answer to the riddle of whether there is life on Mars. Even after a concentrated study of the planet over the past decade or so the answer is still not forth— coming.

Schiaparelli in 1890 found faint linear markings on the
Martian sur— face which he called
"channels." But Lowell described them as canals

STRANGE SURFACE FEATURES ON PLANET MARS

Most of the following photos were taken by Holger Isenberg, credit MOC 2—112 http://mars-news.de/ from the out of print booklet entitled: THE SAND WHALES OF MARS.

So far as the actual pictures are concerned, they come from the current Mars Orbiter Camera (MOC) in polar orbit around Mars, built and owned by Dr. Michael Mali n, who is able to hold on to the down— loaded images for 6 months if he wishes, which means forever. We do not think that Dr. Malin owns Mars. Effectively, Dr. Malin and NASA have said — here are our blurry images; make of them what you will. The real answers to the riddles of Mars will only come when humanity is good and ready to go there, and having seen, relays the truth back to us 'mere mortals'. The international space station is but a first step, made necessary by the possibility that Earth may one-day wind up looking like Mars.

In a recent book by Anthony Austin, it was pointed' out yet again that we do not have very much time left, whether or not we're hit by some randomly looping erratic made of iron and rock in an earth— crossing orbit, because we're rapidly   'using up' the planet

through apparently unstoppable population explosion. In 'The Dragon's Tail it is stated that we have no more than 115 years left, whatever we do, as sure as night follows day. The conclusion reached is that we are headed for a new Ice Age. Once again, Mars and Earth will look much the same.

The Maldekians attacked the lunar surface where the
Reptilians guarded their Earth outpost from invasion. The Maldekians also bombarded Atlantis and Lemur ia with laser weapons. The dinosaurs were wiped out.

Additionally, the Martians also attacked the Reptilians from space since they, too, were searching for a Reptilian—free environment in which to live. This might be considered the real First World War on this planet. It was a real mess!

References: 'BLUE BLOOD, TRUE BLOOD' by
Stewart A. Swerdlow, 2003. Pages 20—22, Expansion Publishing Company, Inc. P.O. Box 12, Saint Joseph MI 49085.

In the mid—1980s one of New York's daily papers, Newsday, reported that a Soviet space probe penetrated the cloud layer of Venus and photographed 7 white domes the size of small cities, all in a row. Please note the photo of domes also, in photo of Venera 15 mission to Venus.

The Reptilians drove a large, hollowed out object into Earth's orbit to begin the colonization process. This object is now called the Moon. The Moon faces the Earth in the same position all of the time, leaving one side in complete darkness. A sonic resonance sent to the surface of the Moon makes a pinging noise like a hollow object The Moon is definitely hollow!

In the meantime, the Martians were now living underground with their hostile Maldekian guests.

Something had to be done quickly to prevent them from destroying one another. So, the Martians petitioned the Galactic Federation to remove the Maldekian refugees to another planet.

Mien the at lans arrived on the Earth, they colonized what became known as
Atlantis. Their continent stretched from what is now the Caribbean Basin to the Azores and Canary Islands, as well as several small island chains reaching up to what is now the East Coast of the U.S., including Montauk Point.

The industrious Atlanteans rapidly grew to a large, prospering civilization needing more territory. The dinosaur population was rapidly increasing and becoming dangerous to the human colonists. The Atlanteans began destroying the dinosaurs to protect themselves. This did not sit well with the Reptilians. Soon major battles occurred on the Earth between the Lemur i an Reptilians and Atlantean humans.

At the same time, the Maldekian refugees arrived on Earth. They then created a large human colony in what is now the Gobi Desert, northern India, Sumer and other parts of Asia. to its target. All of the technology was obtained by the beings from Sirius A.

In this way, they hurled a huge ice comet aimed at Mars and Maldek. The Reptilians, not being very technologically oriented, miscalculated the trajectory. The pull of the gigantic gas planet, Jupiter, pulled the comet off course. The ice comet then headed directly for Maldek. The citizens of that planet asked the Martians for help. Even though they were at odds with each other, they allowed some of the Maldekians to move to the Martian underground. The comet came so close to Maldek that the planet got caught between the gravitational pull of Jupiter, Mars and the comet. This caused the planet to explode, leaving an asteroid belt between Mars and
Jupiter.
The explosion pushed the ice comet close enough to Mars to rip the atmosphere off that planet, leaving only an extremely thin atmosphere. The explosion also pulled Mars further away from the sun.

The comet then continued on toward the Earth. The heat of the sun and the gravitational pull between the 2 globes forced the watery atmosphere of the Earth to polarize. This polarization pulled most of the ice from the comet to the polar regions of the Earth, thus covering most openings to the inner Earth, while at the same time exposing huge land masses for the first time.

The comet then switched places with Earth, taking up the second orbit from the sun, becoming planet Venus. The heat of the sun melted the ice on the comet, creating a cloudy covering to this new planet,

The Earth was pushed out to the third orbit occupying the previous position held by Mars. The Earth was now ready to be colonized. Most of the surviving amphibians were transported to a new home on Neptune. Some stayed in the newly formed oceans.

The Reptilians who were inside the hollow comet, now Venus, came to the surface of this new world. They built 7 domed cities, one for each

The Great Pyramid is part of a protective solar system grid, linking the Moon and Mars monuments together to produce a force field to repel invaders. The Great Pyramid is also connected to other points on the Earth such as Stonehenge, a submerged Atlantean crystal, Tiahuanaco, Ayers Rock, and the White Pyramid in western China. Together, they form an energy containment field similar to an electric fence. The HAARP project in Alaska taps into this.

THE REPTILIAN AGENDA

The Reptilian agenda was, and is, to seek out the human refugees for destruction or assimilation, and to use their blood and hormones for sustenance.

The remnant Lyraens who colonized other planets formed an alliance against the constant Reptilian attacks. This was called the Galactic Federation, comprised of 110 different colonies. Together, they managed to repel the Reptilian attacks.

There were 3 primary groups who did not join the

Federation. One group was the
Atlans, located on a Pleiadian planet. The Pleiades actually consist of 32 planets orbiting 7 stars. At that time there were 16 different colonies of Lyraen descent throughout the Pleiades. These colonists all wanted to oust the renegade at lans because they remained independent and did not assist their human cousins.

The other 2 groups were the Martians and Maldekians, who were already at odds with each other. For this reason, the Reptilians turned their attention toward this solar system with its 2 human colonies. In the Reptilian's estimation, it would be easy to divide and conquer.

The Reptilians love to use comets and asteroids as weapons and ships, using them to travel through the stars, First, they create a small black hole as a propulsion system that pulls the larger planetoid towards its destination. When used as a weapon, they use a particle beam accelerator to create a blast that hurls the comet or asteroid in the high astral levels, there is an ether ic race of Lion Beings who have wings and violet eyes. This race is called Ari. Ari is the old Hebrew word for lion. Their frequency is more powerful than the dolphin frequency. The Ari created the Ohalu Council that governs the Sirius A star system.

The 'Kilroti' beings were generated by mixing the genetics of the Syrians with the energy of the Ari. This is what was brought to ancient Egypt. As the non—physical energy descended into physical reality, DNA formed that could be used to create corporeal life.

This was then mixed with human and wild lion DNA to form the common house cat found here on Earth. The cat was given to every home in ancient Egypt and programmed to leave at night to report back to their alien controllers. This is why cats to this day have the urge to go out at night. This also explains their aloof nature.

The Syrians incorporated worship of the cat idol into the Egyptian religion to ensure the perpetuity of this method of spying. The Sir i an:

also built the Sphynx as a symbolic reminder of the blending of human genetics with lion frequency. This was a way to energetically bind future civilization to the Syrians. The Sphynx was designed to face the morning star Sirius A every day. The face on the Sphynx is identical to the face on the Mars monument that looks down to the Earth at the Sphynx.

Syrian technology built the complexes at the Cydonia plateau on Mars, upon the arrival of the first Lyraen refugees. The new Martians were unaware of the close Syrian connection to the Reptilians.

The original pyramids were built about 73,575 years ago. The original pyramids, built after the destruction of Atlantis, were energy points. They were the same shapes underground as above, making them into octahedrons. At their center is a tetrahedron. This master shape is the archetype symbol for God—Mind totality.

THE REPTILIANS UNDERGROUND ON EARTH AND   ON OTHER PLANETS

The Chinese Reptilian religion spread across Eastern Asia, while the Sumerian version ran through Central and Western Asia. The spread of these religions was intentionally controlled from the underground Reptilian population, primarily centered under Tibet. These reptilians were aided by beings from Sirius B who developed the Buddhist philosophies, as well as a group of renegade Lyraens trying to reproduce a Lyraen civilization under Reptilian control. Strange bedfellows!

At the same time in India, the Lemurian Reptilian refugees created a caste system that vas a direct replica of the Reptilian hierarchy, from the lowly workers/ untouchables to the Brahmin/winged ones. This Indian/ Reptilian culture remained localized, writing the ancient Rig— Vedas and building temples to the various Reptilian gods.

Meanwhile, the Egyptians, who were Atlantean/ Lyraen refugees, were in the process of building a new civilization. The beings from Sirius A helped them, as they were a major factor in the interactions of Atlantis.

In Egypt, the Reptilian gods were known as Osiris and Isis. The Egyptian panacea of gods included a large variety of hybrid creations, half—human, half—animal. This was reminiscent of the Atlantean hybrid experiments that found their way into Egyptian culture, and was promoted by the Syrians who were preparing that culture for a Reptilian takeover.

In the Sirius A star system, the main world is called Khoom. The old name for Egypt is Khem. There is also a correlation with Mexico. Some researchers say that the Bay of Campeche translates to the Bay of Old Egypt, indicating a connection between Egypt and the Yucatan Peninsula. This is not so. Syrians who interacted with Atlantis named this area after their home world, then carried the name on to the new refugee culture in Egypt climates and an oxygen— rich atmosphere. The gravity on Maldek was denser than Mars, so those people developed a thicker frame and a more aggressive attitude.

Eventually, skirmishes developed between the occupants of the two planets. Mars was rich in resources. The people of Maldek thought that they deserved these resources for survival. The Martians asked the beings of Sirius A, from planet Khoom, for defense technology to shield their planet from attack, not only from the Reptilians, but from their humanoid neighbors and cousins. The Syrians are known throughout the galaxy as merchants of technology. They have the best interdimensional technologies, even sharing it with the Reptilians. So, the Syrians developed and created a defense mechanism located in the underground cities of Mars.

Mars is a hollow planet, as are Earth, Jupiter and most of the other planets in our solar system. Planets created with materials ejected from a star have hollow interiors. As a molten ball is thrown from the star and starts spinning away, it begins to cool. The centrifugal force of the globe spinning and moving at great speed pushes the molten interior to the sides, forming the crust of the planet.

This, in turn, forces hot gases out of the poles to form openings at both ends. The molten core and gases that remain get trapped between the hollow interior and the plates below the crust of the globe. These are pushed out periodically in the form of volcanic activity.

The nexus point on any such globe or planet is always at the 19th parallel of the planet. It is evident on Earth by the Hawaiian volcanoes; by the Mons volcano on Mars located at the 19th parallel; and at the red spot on Jupiter, also at the 19th parallel.

The geometry built into the monuments on Mars and Earth by the Syrians and Lyraens explains about the 19th parallel through its geometric equations and measurements. This sacred geometry is also replicated and contained within the Giza plateau in Egypt.

EACH MARS, MOON + VENUS HOLLOWED OUT PLANETS WITH EXTRATERRESTRIAL

BIOLOGICAL ENTITIES LIVING THERE

As indicated by the evidence of several other scientists, planet Earth has been invaded by a few extraterrestrial biological entities including the 'REPTILIANS', PLEIADEANS, LYRAENS, SYRIANS.

The Lyraens made their homes here, about 1 billion years ago. They did not have a defense system in place when they were attacked by the Reptilians, also known as the Dracos from Alpha Draconis. The survivors of the Lyraens dispersed to other locations throughout the Milky Way galaxy.

These survivors went to Orion, Tau Ceti, Epsilon Eridani, Antaries, Alpha Centauri, Pleiades, Procyon, Barnard Star, Arcturus and many other solar systems. In our solar system, the refugees colonized planet Mars. At that time, Mars was the third planet in the solar system. Maldek was the fourth planet and was also colonized.

The Reptilians did not have much psychic ability. Often, when the Reptilians came to a world for occupation, the Lyraens offered a group of psychic, red haired people to appease them for a while. This then eventually degenerated into sacrificial practices to appease the blood—sucking demons.

Planet Earth in those days was a water world in second orbit from the sun. There was little land above the surface. The only intelligent inhabitants were an amphibian race that was completely without technology. The atmosphere of the Earth was mostly liquid. It definitely could not sustain any type of human life forms.

The Lyraen descendants developed their own cultures over the eons of time. Their genetics manifested differently as a result of the mind— patterns of each of the colonies. For instance, Mars and Maldek were similar to the current Earth environment, with warm to temperate

Structures such as buildings or one's own home are created by materialization through psychokinetic abilities.

This gives rise to great variations in design or architecture. The more advanced cities are on mountain tops, and the least developed in the valleys or on the flat plains. As we have mentioned these lower levels of consciousness are well above those of Earthlings, but nevertheless many of these beings still have ties with materialistic past existences and feel the need for corresponding experiences, such as eating.

Buildings may be enormous with extremely spacious interiors, although average height of the Venusian is not much greater than Earthman. Cities may present a beautiful array of domes, minarets and spires, thoroughfares adorned with plants, and blossoms befitting a horti— culturist's dream.

The construction, positioning and shapes of certain centers are architecturally designed to filter and regulate influx of cosmic energies for meditative purposes and communication with other worlds or universes. In particular, Venus is noted for its dispensation and ad— ministration of healing energies and consequently is referred to as the mother planet for the Solar System and is the most spiritually advanced planet in our system.

Noel Huntley, " Escape from the Universe" 1985 out of print now dimensions. This places their existence outside the material 3D band of reactionary energies and thus not vulnerable to the severe physical or 3D environmental conditions on planet Venus. However, all levels and dimensions interact and there will be correspondences between the higher level atmosphere and the physical one. But even at the physical level on Venus there is consistency, good behavior of weather conditions and no snowy regions owing to the planet's axis not being tilted, as is Earth's, creating the seasons.

In higher dimensional worlds, such as this, lower forms of life exist but there is no predatory survival or survival of the fittest. Nourishment is ingested by means of prana (cosmic energy) in respiration in the relatively lower forms, but in the more four— dimensional energy forms beyond the atomic configurations of the upper echelon of the Venusian t s development nourishment is in the oscillator) contact or resonance with the Infinite.

These higher—frequency civilizations on Venus enjoy an environment of iridescent glowing beauty and brilliance of color indescribable to Earthlings:
radiant opalescence of oceans, rich bounteous vegetation, crystal mountains, and harmonious animal, bird and insect life.

There are societies which are at different levels of the scale of consciousness and exist separately in different cities. In general, however, inhabitants will encounter crystal cities, fairy—like castles, forest glades, environments lighted by the perpetual radiance of the Infinite, kaleidoscopic and multi-varigated forms of birds and butterflies, rainbow hues reflecting sparkling shafts of radiant light, and an invigorating " atmosphere" free from pollutant energies. The lowest form of man on Venus is said to correspond to the wise adept or guru on Earth, and even a child's intelligence would outshine any bright human. To become an inhabitant of Venus and there is no birth as man understands it a person will telepathically, and from their standpoint, reach out spiritually, to a plane or place harmoniously suitable, then seek out a family and communicate their wishes. The family group, by means of psychokinesis builds a body for the new individual who enters into it. The participants become the parents who may have to replenish the body structure from time to time until the new spirit makes the necessary adjustments.

These people have developed their consciousness to a high level of sanity, intelligence and love for all creation. Their bodies are described as about an average of 5 feet 6 inches and are similar in appearance to our Chinese race.

Advanced technology enables children to learn mainly by video with the addition of sleep learning in which a ray transfers knowledge sub— consciously. Birth has an alternative process to the natural one. The fetus can be developed externally in special vessels if the mother wishes. This has advantages by both the presence of visual communication from the mother, and also the enhanced learning process of the infant achieved by scientific means for absorbing information relating to the infant, such as potential abilities, past—life experiences and the present goals for this coming life.

The Martians can travel between galaxies in their spacecrafts but even with their advanced technology enabling them to travel hyperspatially it takes them years. Such ships as theirs may contain as many as 5000 passengers, although they refer to encounters with city ships from other civilizations containing up to one million people.

Martians, it is explained, did visit Earth and set up a colony in the Gobi Desert.
However, the primitives of Earth were found to be so backward and hostile that this Martian settlement was abandoned. Thus we see then that Mars does have a very advanced civilization but seemingly not any longer in this dimension.

Planet Venus presents a similar situation but is even more spiritually advanced and definitely not in our dimension.

The surface of this planet has always caused a mystery to astronomers, with its perpetual veil of clouds. The information given in Ernest Norman's book 'The Voice of Venus' is that this blanket of vapor is not a natural occlusion. The surface has been shielded by Venusians deliberately by means of high—frequency energy, with aetheric condensation forming an envelope around the planet; the purpose being to prevent man's confusion in advancing science's possible detection of inexplicable elements and strange radiations.

There are, in fact, magnificent civilizations on Venus which exist at higher vibrations of the atomic vortex and correspondingly greater in our dimensions. This question is not made clear, but this is supported by the fact that out— of— body visits (overshadowed by the spiritual hierarchy) have been made by Unarius channels from whence came much of the information, and also other sources such as the fact that the spiritual hierarchy directed the massive negative thought forms remaining after the destruction of Maldek a civilization on a planet beyond Mars of which the fragments now are the scientifically unexplained asteroid belt between Mars and Jupiter and attached it to Mars to keep it away from Earth.

If Mars had been inhabited this decision would not have been made.
Nevertheless, this is supposed to be the reason why Mars is known as the war— waging planet of somewhat hostile vibrations.

While on this interesting point another reason for Mars being associated with war is that prior to the destruction of Lemur i a, Lemurians, using their space fleet, launched an all—out attack on Mars.

The inhabitants were prepared and using a surprise strategy defeated the Earthmen. However, propaganda continued to the effect that, in fact, it was Mars which attacked Lemur i a, thus giving Mars the reputation of being hostile.

In our context then it appears that this Mars civilization is no longer in our physical dimension. This would also tend to be supported by the fact that this civilization has continued for hundreds of thousands of years without destruction, implying great spiritual development has been taking place. As the frequency of consciousness rises so the environment eventually passes beyond physical perception. Nevertheless, the existence of the Martian civilization is still quite physical and involving corresponding material needs of man. The "canals" of Mars are supposed to be the surface ground effects (or after effects) of the great interconnecting transport system tunnel— ling from one city to another.

A typical city was about one mile in diameter consisting of beautiful buildings and sculptured gardens, fountains, fruit trees, statues, murals and, in general, a horticulturist's paradise. The enclosed roof area of these underground cities appears as a sky with simulated twinkling stars even a "moon" is revealed periodically during the night cycle, and for day—time, artificial sunlight is used. Transport and proposed that their extensive and geometric construction implied the existence of "rare intelligence". Many further speculations and observations on Mars led scientists to believe that life on Mars was a reality.

In 1965 Mariner IV reached the vicinity of Mars' surface and transmitted pictures back to Earth. Fuzzy pictures revealed a barren planet with lunar—like craters indicating a surface unchanged in a millennium. In 1969 two spacecraft were sent to Mars revealing still more austere and hostile conditions and the unlikelihood of life. Undaunted and far from satisfied, scientists launched Mariner IX which went into orbit around Mars in 1971, but on the side of Mars not previously seen four huge volcanoes were apparent which dwarfed our highest mountains.

Furthermore, a "Grand Canyon" was revealed some 3000 miles long and 15-20,000 feet deep.

Many geologists are convinced that the 30 mile— wide canals first seen by Mariner IX were formed by giant floods caused either by meteors impacting and melting ice under the surface of Mars, or volcanic activity. However, an immense amount of water movement is required to explain the massive rifts and plains.

 From all the information obtained by the space probes scientists are little nearer to understanding Mars and, in fact, the geological material has seemingly posed even greater questions.
Now, the so—called planetary "Nova" which destroyed Lemuria affected Mars by dehydrating and devastating its surface. Nevertheless, the civilization even then was sufficiently advanced and in

communication with the spiritual hierarchy to receive prior information of the intruding the body and prepare by building cities underground.

The nova was actually a manifestation in the third dimension resulting from higher— dimensional energies of a positive kind directed at the negativity enshrouding planet Earth in order to cancel it. The positive and negative energy interaction expressed itself in the apparent event of a so—called planetary body entering the Solar System.

Thus the Mars' civilization continued, and according to Norman's book the civilization is still in existence, underground! Taking into account other data, the indications though are that it is no longer.

As the consciousness of man expands, his vibrations of mind and body rise in frequency. At the present time this is occurring with an in— creasing number of people on the planet and consequently this will influence the vibration of the planet towards greater frequencies. Any sudden changes though, even of a positive nature such as the effect of the Christ consciousness through Jesus could upset the balance of the energies of the planet. At the time of Jesus, the huge influx of negative forces (which were allowed by the Creator or spiritual hierarchy) produced a huge negative factor which balanced out the positive one from Christ.

At the present time, the balancing element is being provided by the fact that the negative elements on Earth are increasing (just as the positive ones are increasing). Bad music and art are becoming worse, satanic cults are expanding and drugs and disharmony are becoming more widespread in these lower ranges of consciousness' vibrations. During the thousand— year period planet Earth will gradually fade from the 3D space—time matrix as its lower harmonics withdraw up the frequency scale and hierarchy of levels of consciousness.

ADVANCED LIFE ON MARS AND VENUS

Cosmic visionary Ruth Norman of the Unarius
Foundation informs us in her book 'Mars Underground Cities Discovered' that Mars was the victim of a huge disturbance some 160,000 years ago, described as a Nova blue involving the entry into our solar system of a body passing close to Mars, colliding with Earth, and destroying the continent Lemur i a. We shall return to this but in the meantime let us take a look at contemporary science's information on Mars.

Scientists continue their age—long quest for the answer to the riddle of whether there is life on Mars. Even after a concentrated study of the planet over the past decade or so the answer is still not forth— coming.

Schiaparelli in 1890 found faint linear markings on the Martian surface which he called t' channels. " But Lowell described them as canals.

STRANGE SURFACE FEATURES ON PLANET MARS

Most of the following photos were taken by Holger Isenberg, credit MOC 2—112 http://mars—news. de, from the out of print booklet entitled: THE SAND WHALES OF MARS.

So far as the actual pictures are concerned, they come from the current Mars Orbiter Camera (MOC) in polar orbit around Mars, built and owned by Dr. Michael Mali n, who is able to hold on to the down— loaded images for 6 months if he wishes, which means forever. We do not think that Dr. Malin owns Mars. Effectively, Dr. Malin and NASA have said — here are our blurry images; make of them what you will.

RINGS OF FIRE AROUND EARTH

The real answers to the riddles of Mars will on one day come when humanity is good and ready to go there, and having seen, relays the truth back to us 'mere mortals' The international space station is but a first step, made necessary by the possibility that Earth may one-day wind up looking like Mars.

In a recent book by Anthony Austin, it was pointed out yet again that we do not have very much time left, whether or not we 're hit by some randomly looping erratic made of iron and rock in an earth— crossing orbit, because we 're rapidly 'using up' the planet through apparently unstoppable population explosion. In 'The Dragon's Tail ' it is stated that we have no more than 115 years left, whatever we do, as sure as night follows day. The conclusion reached is that we are headed for a new Ice Age. Once again, Mars and Earth will look much the same.

Doomsday theorists have had their hands full recently, correlating every prediction from Nostradamus to the Mayan calendar with the current spate of earthquakes and volcanic activities. Since 2006 Haiti, Chile, China, Japan, the United States, and several other countries have experienced temblors measuring very high on the Richter Scale to cause very widespread destruction and, in the cases of Haiti, Pacific islands, Chile, and China, massive losses of life. Add in the chaos caused by new volcanic eruptions in Ecuador, Italy, Japan, Iceland, and in the United States, and one quickly understands how the various theorists of the conspiracy world may believe that the sky is falling and more worldwide disasters are going to happen from now until the end of year 2013!! 3 THEORIES

Online journalist A. Waite's article "Atomic Destruction Without the Bombs, " in the defunct blog Outrider 12, reported a link between the violent upheavals on Earth with unusual solar activity. Waite notes "On 10 April 2010, the Solar and Heliospheric Observatory (SOHO) satellite recorded the death knell of a previously unknown comet as it was consumed by the sun. Researchers for NASA and various other astronomical organizations around the globe claimed this is not a rare phenomenon, only that such an event has never been recorded. " Waite points out other disruptions in the normal 11— year sunspot cycle, including records from the Aickman—Barker Science Academy of similar coronal mass ejections (CME) which preceded the catastrophic eruption of Krakatoa, east of Java, in August of 1883, and of the Lowell Observatory's notations of peculiar solar flares seen prior to the extraordinary Tunguska blast over Siberia in 1908, The writer quotes the scientific evidence of F. J. Whipple, supposing the latter explosion to have resulted from the impact of a comet with Earth. Returning to the present, Waite describes how, 3 days after the 10 April meeting of the sun with the unknown comet, " SOHO also recorded an enormous solar prominence reaching a size of 500,000 miles, stretching nearly half— way across the surface of the sun before erupting out into space, an event she claims aligns with the worsening of the volcanic eruption of Eyafjallaj okull, in Iceland, the very next day.

Amateur astronomer and professor emeritus, Dr. Gordon Gregor states that the doom criers are simply "misguided, " but argues that re— searchers like Waite could find more accurate information by gazing into a crystal ball. Dr. Gregor does not deny the wealth of evidence collected by the scientific institutes mentioned in the journal article, but he cautions his colleagues to make informed suppositions based on "fact, not coincidence."

Dr. Gregor sees such activity as cyclical in nature," agreeing with the late Michael Crichton, that even "so— called 'Global Warning' is nothing more than a repetitive phenomenon of the Earth's heating up and cooling off, which the professor claims will continue to occur ad infinitum into the future until the sun super novas or "the little green men whisk us away a few at a time in order to blend us into pate or have us as human barbeque" (see the Twilight Zone episode "To Serve Man, 'J 1962).

The professor also mentions a fact corroborated by recent data from NASA concerning the increase in surface temperature of not only Earth, but of Venus and Mars. Dr. Gregor and several of his former colleagues believe the recent volcanic activity on the moons of Jupiter and Saturn may also be a result of this solar cycle.

The research team of historian Brian Leno and investigator S. H. Urban propose what seems, at first, the least likely of causes for these recent disasters, yet, upon further consideration, may well be the most terrifying possibility. Leno and Urban lay blame for the volcanic eruptions, earthquake devastation, and even the escalating public un— rest in the world to a single, man— made cause the European Organization for Nuclear Research's (CERN) Large Hadron Collider (LHC). Leno scoffs at the reported numbers for the LHC's output. "The scientists there are fairly straight—faced about their operation, maintaining their evil spiel about how they hope to ramp up the LHC to the point where they can get their proposed seven trillion electron—volts by colliding single protons. It's a good cover for their actual aims.
 "Urban agrees and claims that his partner's theoretical possibilities led him to investigate "What's really going on there; the creation of a wormhole that can be turned on and off. Such a device could be used by the most powerful governments for the purpose of time dilation.

When asked for evidence to support the enormity of their claim, Leno revealed several notebooks filled with intricate mathematical calculations relating to the Large Hadron Collider's specifications and the actual limits (kept secret thus far) of just how much power can be generated. "That 17—mile monster can easily produce 3 micro joules per particle. The bastards could solve the world energy crisis in a week, but they're too busy perfecting the Time Tunnel to bother. These real criminals have already wasted billions of Euro$ on this project that will never work based upon their negative and destructive aims!

Political opinions aside, Urban's research in the surrounding region of Switzerland and France has turned up numerous physical changes in the terrain, such as the upheaval of long stretches of roadway, sections of buildings and bridges which have shifted out of true, and in 3 separate locations, evidence of rivers and waterways that had been slewed from their former paths. In one location, near the Franco— Swiss border, a small village was temporarily flooded by such an alteration. Urban also points out odder phenomenon, such as individuals who have disappeared and reappeared before startled witnesses (similar to reports associated with the so— called "Philadelphia Experiment" of 1943), a disproportionate number of cases of dementia and Alzheimer's Disease (particularly alarming in Meyrin which lies just outside one edge of the LHC's largest ring), and a general consensus of malaise and depression throughout the region. These geological, physical, and mental aberrations are indicative, says Urban, of "A warping of the physical reality we know, and one that could easily explain the un— usual stresses occurring around the planet as the earth tries to counter for this displacement of its own relative position in the fabric of space—time. " Both researchers note the strange level of geological activity and civil unrest since 2008 when the LHC was first tested. Leno also notes an interesting coincidence in one of the proposed dates for the LHC's search for the Higgs boson particle 21 December 2012.

"Maybe," says Leno, "the Mayans had it right after all. This author maintains that the Mayans did not have it right after all, as there is no beginning and no ending in "GOD'S INTERDIMENSIONAL UNIVERSAL EXISTENCE".

THE BIG BANG THEORY IS ALSO WRONG, DEAR FRIENDS!

THE OTHER RING

Although the 3 previous theories are equally compelling, each one bows to the statistical evidence seen in all—too earthly phenomenon the Pacific Ring of Fire, one of the largest areas of tectonic plate coastlines from Micronesia up around Asia and back around and down the coasts of North and South America. The region accounts for over 90% of the world's seismic and volcanic action and may well account for the majority of current activity. A search for Alaska alone turns up 50 volcanic eruptions since 1760, including Mt. Redoubt in 2009. For old school prognosticators who once predicted that California would eventually slide off into the ocean due to a violent earthquake, the more likely scenario is that Mt. Rainier will erupt, destroying Seattle and its environs.

Of course, not even the Chicken Littles want to be right about such phenomenon. Few, if any, would actually wish to stand amid the rubble saying "I told you so." Yet, no matter whose idea one chooses to espouse, the facts point to major upheavals in the -world. If Asenath Waite's theory holds, then we might soon see a tapering off of such violent actions as the sun calms into a more normal rhythm of sun spot activity. Dr. Gregor's best and worst case scenarios relating to cyclical changes include a leveling off of world temperature and eventual cooling as we saw in the 1970s and early 1980s, or plunge in temperatures resulting in another Little Ice Age like those occurring between 1650 and 1850. Worst still, the entire planet could be warped by the actions of the Large Hadron Collider and torn asunder by the stress resulting in opposing gravitational forces.

Some pundits claim that we've had it too good for too long and have grown lax in our abilities to cope with these geological and climatological changes.

Such suppositions do not bode well for the outlook of the world's societies. But perhaps the greatest conspiracy here isn't some unknown technological horror about which our governments refuse to share information. The most frightening fact of all may be that we 're currently doing this to ourselves. What if we have ways to stop it yet continue to destroy the only home we have in favor of acquiring some granule of knowledge, some paltry fact that will benefit only the smallest quotient of the world's population??

Tesla's dynamic theory of gravity was originally revealed in this author's book back in 2006/2007, entitled 'GRAVITY, MATTER + SPACE

TRAVEL'. Tesla's main criticism of Einstein's invalid theories of

'relativity' had to do with Einstein's idea of the curvature of space: "On a body as large as the sun, it would be impossible to project a disturbance of this kind (e.g. radio broadcasts) to any considerable distance except along the surface. It might be inferred that am alluding to the curvature of space supposed to exist according to the teachings of 'relativity', but nothing could be further from my mind. I hold that space cannot be curved, for the simple reason that it can have no properties. To say that in the presence of large bodies space becomes curved is equivalent to stating that something can act upon nothing, I, for one, refuse to subscribe to such a view. Nikola 'Pioneer Radio Engineer Gives
Views on Power, " New York Herald Tribune,
9/11/1932

George Gamow, one of the founders of quantum physics, states that, in the
1920's, in the process of measuring the rate of electron spin, Goudsmit and Uhlenbeck discovered that the rate was 1 .37 times the speed of light! AS Gamow tells us, this did not violate any principle in quantum physics; what it violated was Einstein's theory of relativity.

Tesla, however, never abandoned the ether. Neither did Einstein or Lorentz, but nobody writes about it; and the ether has remained the elephant in the room for more than a century.

Now let's return to Tesla and his quote above. Starting with the first sentence about directing a disturbance around the surface of a large body, what Tesla is actually talking about here is 2 related concepts. The first is the ground connection in radio transmission. The second is why there is the need for the ground connection. What Tesla had done, throughout his life, was carefully obscure the reason why radio broadcasts follow the curvature of the Earth — follow the ground connection. The answer has to do with the single sentence he revealed to Alsop in 1934; namely that the Sun was absorbing more energy than it was radiating.

This is Tesla's dynamic theory of gravity. All matter is constantly absorbing ether all the time at the tachyon ic speed of 1.37 times the speed of light. This is the world of interdimensional ether.

By God 's nature, the ether exists in a realm that transcends the speed of light.

That is why radio broadcasts follow the curvature of the earth. As the wave propagates, it is pushed down to the earth by gravity, by this constant influx of ether. So, the reason light particles bend around stars and planetary bodies is not because space is curved, but because these photons are being affected by this constant influx of ether. This theory further speculates that photons are not really mass less, their mass would be equivalent to Planck's constant, a tiny factor which Planck had to add to all his calculations to make them work out. If photons have energy, and if energy and mass are equivalent, then by definition, photons must have some mass. Gravity is the influx of ether by the elementary particles, it is the process that gives matter its "mass". This process occurs at 1 .37 times the speed of light. As each elementary particle absorbs ether, two things happen — the process allows or helps the particle to continue spinning and, simultaneously, the energy is converted into electro— magnetism. Either comes in, causes electrons and other elementary particles to spin, and in that process, atoms retain their integrity and convert the constant influx into the electromagnetism. Inter— stellar speed of light is far beyond the 1 .37 times the speed of light for our solar system... The intergalactic speed of light is truly our GOD'S INTERDIMENSIONÄL, GRAVITATIONAL ESSENCE, AS THERE IS NO

BEGINNING AND NO ENDING IN GOD'S UNIVERSES... GOD BLESS YOU ALL, MY DEAR FRIENDS!

FjNRLCY gh+0$

is lvor

UR ᴴ

cobs VERSES EXPFr/VD-)Æ4

REFER 70 GRAVY, /vlA'TTé4QsPhéE

U

Bock/BY

PRcF. bß. 'bf-os J. PETERMANN ─ FOR F-VQ-
,L/ER

ENERGY 8 TESLA

Fohat — Mind — Life Force — Eloptic Energy — Thought Force E.S.P., have tried to express what little is understandable about an effect that manifests through all living things.

Even those few who do understand it, would want to attribute its universal essence to the Creative Spirit which makes all life eternal. We have discovered that electricity and magnetism are the right and left hands of God, so to speak. Without electro—magnetic energies a Sun, or Planet, or a Solar System; an atom, or structures of the atom, living bodies, ants, or people, flowers, birds, or fish, nothing can be without these invisible forces of electricity and magnetism. In polarity response they both interchange with each other and one cannot exist without the other.

The mysterious energy that makes electricity and magnetism work, is perpendicular to both the electric and magnetic patterns, and establishes a third plane of reference which we call a "time zone. The time "zone" is not detectable, or measurable, with standard electric or magnetic instruments. It responds to thought and can orient the magnetic pattern out of perpendicularity to the electric pattern. If the magnetic field exceeds 87 0/o and coincides to the electric wave pattern, disintegration results in matter and it becomes energy. This occurs with no sound, or heat, or explosion, but will be accompanied by brilliant light emissions. Teleportation can be accomplished by this effect. Pure elements of matter can be created out of energy in the spectrum frequency of any element with controlled polarity reversal in the electric flow.

To our understanding this "zone" is universal mind which creates reciprocal actions between electric and magnetic effects. This "zone" being responsive to thought, makes it possible to understand how all matter from galaxies, to people and atoms, can be created by a Universal Mind, or Cosmic God, which is everywhere. This could explain the old expression, "never let your one hand know what the other hand is doing, " because the other hand would set up an equal and opposite effect. This also would explain how a Supreme Infinite Intelligence could create matter and the physical universe out of the unseen energy. Mind had to come before matter, otherwise how can one explain the order of celestial bodies in unlimited magnitude down to the complete order in atoms.

Should one feel that they are infringing on God to try and understand his creations? Should one feel that it is "off bounds" to try to find out how the perfect laws of the universe operate life in matter? There is no "tree of knowledge" Knowledge is in finding out what makes a tree a tree. Wasn't man supposed to have "dominion over all things? Wasn't man given a "right of free choice" by God?

300 trillion cells in one adult human body perform an interchange of biological electrical functions in such order that only a perfect Universal Spirit could direct their activities. No human study of effects in physical structure will ever reveal the causes of these effects, unless one works through the spirit in dedication to a beneficent purpose for mankind.

"This other Energy" is the conductor of cell communication between all living things. Cleve Baxter's research proves the communication but it doesn't explain how it works. Cells do not have ears, but they do respond to thought.

The Hieronymus device proved that communication from cell structure in the astronauts in Apollo 8 and 11 had no loss in power from the earth to the moon.

Hieronymus also proved plants would grow in total darkness by bringing "This Other Energy" into the darkness from metal plates exposed to sunlight through wires connected to metal plates over the seeds. Light does not travel th6ugh wires.

Franz Anton Mesmer, who was born in 1734, called this other Energy "animal magnetism." Dr. D'Eslon, Mesmer is chief pupil, formulated the laws of animal magnetism as follows:

"1. Animal magnetism is a universal, continuous fluid, constituting an absolute plenum in nature, and the medium of all mutual influence between interstellar bodies and between the earth and animal bodies. 2. It is the most subtle fluid in nature, capable of flux and of re— flux, of ebb and flow, and of receiving, propagating and continuing all kinds of motion.

The human body has poles and other properties analogous to the magnet.

The action and virtue of animal magnetism may be communicated from one body to another, whether animate or inanimate.

It operates at a great distance, without the intervention of anybody. It is increased and reflected by mirrors, communicated, propagated, and increased by sound, and may be accumulated, concentrated and transported. " Is this increase by sound the reason for singing hymns in the churches?

"This Other Energy" was called "The Mumia" by
Paracelsus. "The Odic Force" by Reichenbach, "The
Nervous Ether" by Richardson, "The X Force by Eeman, "The Prana" by Indian Gurus, and is "Life Force" itself. Form has a lot to do with the manifestation of "This Other Energy. Cones produce a cohesive transmission of this energy for miles. Minerals crystallize in their individual forms because this "Other Energy requires exact order. Pyramids oriented in the planes of North, South, East and West, relative to their faces, will produce vortices and strange reactions in their center. An egg, out of its shell, in a saucer beneath a pyramid will have the albumen harden in less than a week, while another egg in a saucer outside the pyramid remains liquid If a circle of 7 pyramids could be located equally spaced in vacant areas surrounding the Los Angeles basin, the smog could be eliminated. This would require 7 masts 600 feet high and the top, or longest guy wires would have to be on a 52 plane on all sides equivalent to the great pyramid of Gizeh in Egypt. The only disadvantage of this would be malfunction of some aircraft instruments if they flew over the masts. On the other hand, they wouldn't cost anything to operate except the cost of their construction. We have tested pyramids made of welding rods at all 4 corners and they work as well in experiments as if the sides were filled in as with cardboard pyramids.

So much could be accomplished with the use of " This Other Energy" if the orthodoxy of educated scientists would investigate and accept those principles of nature that operate in life itself. No one seems to be able to break through that wall of self—inflicted ego that keeps science within the profit structure of a defunct economy, while humanity goes down to its own destruction in a chemical and radiation disaster.

 It seems that today, the people who know how to do things are stopped in their tracks, by laws and opinions from authorities who don't know how to do things. Maybe this is the destiny of a planet that reaches for the Moon, when it needs something to divert the peoples thinking about the mess it has made on the earth.

A hungry child cannot understand this philosophy. Disease, and poverty, and war are man—made results of wrong thinking over the years. "This Other Energy" can only respond, and reciprocate, according to the fixed laws of the universe. Education can only give from the past. The present must operate on inspiration and faith, or the future is lost.

A past civilization knew how to use this other 'Energy' to levitate anything, including themselves. It manifested in the Kings Chamber of the Great Pyramid to revive dead V.I. P.S. in 3 days, or mummify their bodies if they failed to come back to life. This activity took 28 days, or 1 magnetic month. This is the reason the

Kings Chamber was built off center in the pyramid, in order to be in the proper area of the energy generated.

Modern science, in their two—dimensional electro— magnetic science, cannot accept the fact that a handful of people, today, understand this third plane of reference.

Dr. Hendrick's cell therapy, in Switzerland, proved that organ cells, injected into the body, would only go to the same kind or organ in the body and fortify it.

People are as different as their fingerprints and "This Other Energy" makes them all live. How long are the authorities going to ignore the basic philosophy of life, known only to a few people, when it could add multiplicity to present life spans?

Over 30% more is being spent now, for weapons of war, then was being spent 4 years ago. This 30% spent in the right direction could have already unlocked so much more of the secrets of life and we could have already been on the road to 150-year life spans without aging.

The greatest genius of all time is practically unknown in the annals of American history.

Marconi, Edison, Bell, the Wright brother and Fulton, are known for their contributions to the scientific and industrial world by nearly everyone.

Nikola Tesla was born at Smiljan, Yugoslavia, on July 10, 1856. He was the inventor of alternating current generators and motors, the polyphase system of the transmission of power, and the 4 tuned circuit system which is the basis of radio transmission. He radio controlled a ship model in 1898. By this he laid the basis of all radio tele— mechanics. Tesla was 18 years old when Guglielmo Marconi was born in Marzabotto, Italy in 1874. Tesla preceded Marconi 's radio discoveries by several years. Tesla held 25 patents on electrical motors and generators from
1886, 9 patents on the transmission of electrical power, 6

patents on electric lighting, 17 patents for controllers and high frequency apparatus, 12 patents in radio technics, 5 patents for turbines, and 11 other patents on various devices. Nikola Tesla vas the father of alternating current which made it possible to transmit power over great distances and run the industries of today.

In 1891 Tesla said "There is no subject more captivating, more worthy of study, than nature. To understand this great mechanism, to discover the forces which are active, and the laws which govern them, is the highest aim of the intellect of man."

Tesla said, "The superstitious belief of the ancients, if it existed at all, cannot be taken as a reliable proof of the ignorance, but just how much they knew about electricity can only be conjectured. A curious fact is that the ray or torpedo fish was used by them in electro— therapy. The records, though scanty, are of a nature to fill us with conviction that a few initiated, at least, had a deeper knowledge of amber phenomena. To mention one, Moses was undoubtedly a practical and skillful electrician far in advance of his time. The Bible describes precisely and minutely arrangements constituting a machine in which electricity was generated by friction of air against silk curtains and stored in a box like a condenser. It is very plausible to assume that the sons of Aaron were killed by a high tension discharge, and that the vestal fires of the Romans were electrical. The belt drive must

have been known to engineers of that epoch, and it is difficult to see how the abundant evolution of static electricity could have escaped their notice. Under favorable atmospheric conditions a belt may be transformed into a dynamic generator capable of producing many striki actions. I have lighted incandescent lamps, operated motors and performed numerous other equally interesting experiments with electricity drawn from belts and stored in tin cans.

These last 2 sentences describe exactly the static generator for which Van De Graaf vas later given credit for developing and discovering. Tesla again was excluded from the scientific history and another person was given the credit.

Tesla's reference to Moses' tabernacle, as an   "electro—static generator, is describing our "Integratron" closely as to principle of operation.

It regenerated the body also, as Moses died at   120 years of age, and as it says in Deuteronomy 34: 7, "his eye was not dim, nor his natural force abated."

This is even as our modern life saving technique called
"mouth to mouth resuscitation" was described in the Bible in Second Kings 4: 34, where Elisha went to a child that was pronounced dead; And he went up, and lay upon the child, and put his mouth upon his mouth, and his eyes upon his eyes, and his hands upon his hands: and he stretched himself upon the child; and the flesh of the child waxed warm. "We still have the eye to eye, and hand to hand, contact to learn. It may be the hand to hand contact is like using jumpers from one battery to another to assist it, and the eye to eye could be a form of mental autosuggestion employed.

Many years ago Tesla said, I believe that the telautomatic aerial torpedo will make the large siege gun, on which so much dependence is placed at present, obsolete."

He was right again, as today we have intercontinental ballistic missiles controlled by electronics, which he called teleautomatics. Why does the government bureau of Health, Education and Welfare, state the country is 50,000 doctors short of what is needed in the medical profession? It isn't because we need more doctors, it's because we need less disease. This is an admission that illnesses

are gaining on the medical care of them. Then why not employ Tesla's electro— therapeutic discoveries? Dr. Ruth Drown, Abrams, Hieronymus, and many others, as well as the old "violet ray machine, all of which proved beneficial in so many cases, could be used. Nikola Tesla was financed by J.P. Morgan, George Westinghouse, and several others, who all benefited from their association with him Tesla used himself as a guinea pig in many of his experiments. He said "Soon my efforts were centered upon producing in a small space the most intensive inductive action, and by gradual improvements in the apparatus. I obtained results of a surprising character. For instance, when the end of a heavy bar of iron was thrust within a loop powerfully energized, a few moments were sufficient to raise the bar to a high temperature. Even lumps of other metals were heated as rapidly as though they were placed in a furnace. When a continuous band formed of a sheet of tin was thrust into the loop, the metal was fused instantly, the action being comparable to an explosion, and no wonder, for the frictional losses accumulated in it at the rate of possibly 10 horsepower. Masses of poorly conducting material behaved similarly, and when a high exhausted bulb was pushed into the loop, the glass was heated in a few seconds nearly to the point of melting. When I first observed these astonishing actions, I was interested to study their effects upon living tissues. As may be presumed, I proceed with all the necessary caution, as well I might, for I had evidence that in a turn of only a few inches in diameter an electro— motive force of more than 10,000 volts was produced, and such high pressure would be more than sufficient to generate destructive currents in the tissue. This appeared all the more certain as bodies of comparatively poor conductivity were rapidly heated and partially destroyed. One may imagine my astonishment when I found that could thrust my hand, or any other part of my body within the loop and hold it there with impunity. The only plausible explanation I have so far found is that the tissues are condensers. "

This conforms to Dr. George Lakhovsky's findings "that the cell— organic unit in all living beings, is nothing but an electro—magnetic resonator, capable of emitting and absorbing radiation of a very high frequency."

The human body is matter manifested. All spirit is matter, and all matter is spirit, each manifesting in opposite polarity under different conditions.

Nikola Tesla's inventions are the basic foundation of all electrical power, and without his discoveries, the scientific advances of today would have been impossible.

OUR INHABITED SOLAR SYSTEM

Venus, Mars, and the Moon are important in that they are our closest neighbors, and their habitability can easily be determined. Our space program of the 1960's and 70's probed and photographed them extensively, since these planets were known to be the homes or bases of the UFOs. Life beyond the Earth should have been published as an established fact years ago. 2 things prevented this from becoming our rightful knowledge: the inability, and often the refusal, of the orthodox scientists to go beyond their shared perceptions; and secondly, the absolute secrecy and censorship by the top—level authorities in charge of the space programs worldwide.

But what about the other planets in our solar system? Are they inhabited also? And are their environments Earthlike, or similar enough, that we could travel to them and step out on their surfaces?

George Adamski stated early on that they were inhabited, for he got his information direct from the space people. Textbook scientists and JPL staff say no, based on the old theories, but they do not have any direct information from scientific instruments. To date, scientists have no way of knowing, because we have not probed beneath the atmospheres of the other planets or landed any instruments. And until they understand the true nature of the sun's energy, gravity, and magnetic fields, their speculations will be way off the mark.

All that has been achieved is some preliminary picture— taking from vast distances during momentary fly—bys. For example, the closest pass to Jupiter by a Voyager probe was at a distance of 166,000 miles. Any so—called measurement from that distance would be purely hypothetical. And until scientists can see the UFO evidence here on Earth, and see the true environments of Venus, Mars, and the Moon, there is no reason to accept their opinions regarding the more distant planets. Their speculations are only extensions of theories that have become scientific dogma. There is simply no data which could prove or disprove the theories. In other words, from a purely scientific standpoint, the planetary conditions on either Mercury, Jupiter, or Saturn is still an DC-5 Z open question. The same is true for the planets beyond Saturn.

In order to discuss Mercury, the last of the 4 inner planets of our system, it is important to first have a true perspective of spatial distances. Diagrams in books and magazines are invariably out of pro— portion due to an inadequate page size, and Mercury is perceived as being very close to the Sun. In space the planet's actual distance from the Sun is 34 million miles.

The following is an accurate representation that could be laid out in a large yard or playground. If the Sun were the size of a regulation basketball, the Earth would be the size of a small BB placed 90 feet from the basketball. A smaller BB for Mercury would circle the BB from a distance of 33 feet. To complete this scaled representation, a BB for Venus would be placed at 65 feet, and another for Mars at 132 feet. The actual size of a BB is Try and picture in your mind these sizes and distances for a moment.

Seeing Mercury as a small BB located 33 feet from the basketball, we find that it is not really close when seen at true scale. The planet is usually described as being very similar to the Moon on its surface, and like the Earth on the inside. As early as 1962, radio astronomers found that the night side hemisphere had a warm surface an effect which would require that an atmosphere be present to carry the heat from the day side to the night side. It was also determined from further studies that Mercury had a thin atmosphere.

The methods used were similar to those used for the Moon, (and like the case of the Moon, this means that the density of Mercury's atmosphere will have to be revised upwards when more accurate methods are employed.) When the atmosphere is accurately determined, we will also learn that the surface temperatures on Mercury are quite moderate.

The U.S. space program has sent only one space probe to Mercury. The spacecraft, Mariner 10, was placed into a precise orbit that allowed 3 brief fly—by encounters during a one-year period from March 1974 to March 1975. These one and a half day encounters were the first close— up views of the planet, and a chance to investigate Mercury's magnetic field and its interaction with the solar wind. The mission was also supposed to check up on the "tenuous" atmosphere determined from radio astronomy, but either the Mariner 10 instrumentation was incapable of registering atmospheric effects, or NASA deliberately denied Mercury 's atmosphere, the same as they have denied the Moon's. If Mercury did not have a sufficient atmosphere to offset internal pressures, it would have disintegrated and returned to the elements of space long ago.

The Mariner 10 fly—bys were nights' ide passes, and therefore photographs could not be taken at the close approach, but only from much greater distances during the orbital pass. In fact, it was the worst trajectory for photographing the planet's illuminated surface, and the end results required complicated imaging back at Jet Propulsion

Laboratories. Resolution was compromised also, due to the lack of Congressional funding for a proposed x— band transmitter, designed to top specifications. All of this should be understood when looking at the photos of Mercury taken by Mariner

10.

The spacecraft's orbit brought it to a second pass of Mercury six months later. Its closest approach was 30,000 miles on September 21, 1974. The third and final pass in March, 1975 brought it to within 6,000 miles, but only very small areas were picked up with slightly improved detail. The reason given for the much—lessened opportunity was that there was a failure with one of the Earth—based antenna systems. However, a few high resolution transmissions, showing features of 3 km size, discovered water or river channels in an area now dry. Orthodoxers suggested their standard silly interpretation: the features depicted in the photographs had resulted from the flowing of liquid rock.

Generally, the other pictures look like the front side of our Moon — plains, craters, and flooded basins.

Space scientists were surprised to find a strong magnetic field at Mercury. This field acts as a protective shield to the sun's energy, and thus creates a bow shock in the solar wind, which was also con— firmed by the Mariner probe. Scientists found that the bow shock was comparable and equivalent to the Earth's interaction with the solar wind.

Yet many conclusions came into conflict. The magnetic field at the 034 surface was given as 1% of the Earth's, but it is quite doubtful that our methods can accurately measure things we do not really understand, namely the gravitational and magnetic fields of planetary bodies. It has also been stated that the photographic fly— bys enabled the scientists to confirm the very slow (59 day) rotation rate of

Mercury, due to the fact that the same hemisphere was in view at each of the 3 encounters separated by 6 month intervals. This apparently confirmed an earlier theory, that while circling the sun twice, the planet would rotate 3 times on its axis, giving a ratio of 2:3. But any multiple of 3 would present the same exact face at flyby. Since a strong mag— net ic field is required to create a strong bow shock in the solar wind, and since a strong planetary magnetic field is generally attributed to the fast rotation of a planet, I do not think that we have heard the final answer as to the length of a day on Mercury.

Because the public only sees the end results of a space mission as a series of photographic images in a magazine article, people are very apt to think of a planetary encounter as a live broadcast from space by the space-probe. Actually, the probe is exploring in the dark, and its various instruments have uncertain sensitivities to atmospheric frequencies, radar waves, solar intensity, magnetic fields, and cosmic radiations. Photography is nothing more than limited electronic detection of surface illumination contrasts, which must be interpreted by computer imaging back on Earth. The various instruments send their non— absolute indications millions of miles to Earth in the form of very faint radio signals. These signals are barely discernible from the constant radio noise coming to Earth from space itself, and some— times indistinct from the high frequency noise of the components used to receive the signals.

For example, eight scans by an ultraviolet spectrometer on Mariner 10 were sent back, and initially the signals were interpreted as Mercury having a moon, or satellite. Project scientists thought that they had discovered a moon circling Mercury. The readings were later identified as radio emissions from a 5th magnitude star that was light—years away! And is this not the same type of instrument as the mass spectro— meter
— that device which has always been the instrument for subjective analysis of planetary atmospheres?

Our space probes operate in unknown zones, pick up some type of frequency, from some type of emission, in +ast and complex planetary field. Something registers, and a feeble signal is transmitted to an Earth— based receiver. Schooled in orthodoxy, project scientists try to interpret the signal. Our space probes are exploring and groping in the dark, so to speak.

Their capabilities at present are like looking at a drop of water with a magnifying glass. Compare that with looking at a drop of water with an electron microscope, and you realize what stage of confirmative knowledge we are at in planetary exploration. A Mariner 10 photograph of Mercury is shown in one of the 13 plates, and first indications reveal that its surface does look like the Moon. In fact, its environment may be very much like the Moon. Readers of my last book "Gravity, Matter & Space Travel" are aware that there is a lunar civilization and life—supporting conditions on our satellite. In other words, since our Moon is a bited (contrary to the official position of NASA), we may later find that Mercury shows a similarly inhabited environment, if and when we send out better space probes. From a scientific standpoint, it is still an open question. But just as the electron microscope can reveal a whole new world of activity in the drop of water which the magnifying glass cannot see, so too will our probes discover the actual world of Mercury, when their capabilities are increased 10,000 fold. This same analogy can be applied to the present state of knowledge regarding Jupiter, Saturn and beyond. As stated in my last book,

Ganymede one of the gigantic moons of Jupiter — has humanoid beings living there with their very advanced flying saucers. Many have visited planet Earth in the past years, going back a very long time.

W.R. Drake's researches into classical literature and ancient manuscripts found many references to Mercury. One of the suppressed biblical accounts, the Book of Enoch, describes how Enoch was taken into space (the heavens) and transported to various planets, including Mercury. On this celestial journey he was shown the grand design and motions of the solar system. Enoch's ancient accounts tells of the round shape of the Earth and how our world is inclined on its axis.

That knowledge could only be learned from contact with a space civilization. Enoch went on to be an inspired prophet in the cradle of civilization in Sumeria.

pan in i, a Sanskrit historian and writer, wrote an account of how he and others had been taken on space trips to Mercury and Venus by the friendly extraterrestrials during the 4th century A.D. The famous philosopher, scientist and astronomer of the 18th century, Emmanuel Swedenborg, after publishing great works in scientific discoveries and fundamental theories, intimated that his knowledge had come from telepathic intuition originating with space people from Mercury. In 1866, 3 professional astronomers saw an unexplainable light on the night side half of Mercury. M. K. Jessup attributes this telescopic observation to a brilliant UFO close to the surface of the planet, just as UFOs have been seen over the Moon.

 These are a few of the references to Mercury from the records of history. Intelligent men in the past learned, or otherwise gave indication, that Mercury was a living planet. In our recent history of space visitations, this fact has surfaced again. Officials are more certain with Venus and Mars, but the world has a lot of learning to do. On Earth, mankind laughs at the thought of a divine purpose to creation, denies that there is any other life within light years of our little spot in the galaxy, and contemptuously threatens this most beautiful globe with nuclear bombs. His mind is so divorced from cosmic principle and intuition, that we cannot call it intelligent. Volumes have been written on Jupiter and Saturn, based on nothing but a belief. The theory that these planets are large bodies of swirling gases around a liquid core became a fundamental cornerstone of organized astronomy forever entrenched, never questioned. The psychology of the situation is almost medieval, because the theory is taught as absolute fact by the theologians of science. And if one is paid as a scientist, one has to write and write, with the result that the original belief now expands into volumes of meaningless, nonsensical concepts. Supposition upon supposition, until natural inquiry is stifled, then suffocated.

Johann Goethe, the great German philosopher, once declared that so— called education was expressly designed to make sensible people stupid.

The theories have not changed, but the present discourses on Jupiter and Saturn are based on the picture— taking flybys of the Pioneer and Voyager probes. The more sophisticated cameras were on the hardy

Voyagers, and the closest approach to Jupiter was
166,000 miles by Voyager 1, while Saturn was passed at a 63,000mile distance by Voyager 2 in August 1981. Most of the photography was taken from much greater distances.

The 2 probes' electronic scanning signals were converted into pictures back on Earth, and in the case of Jupiter, computer enhanced into false color. We did get a few close—up images of both of the planets' moons, and a more detailed look at Saturn's rings. The Voyager photos of Jupiter that were released show bright orange and red bands. This was an exaggeration, in order to support the usual description of Jupiter's t' storm—wracked atmosphere." The actual colors as seen from space are soft blues and yellow, and blue—green. This was confirmed by California astronomer

Andrew Young. The best available photographs of Jupiter and Saturn are reproduced in the illustrations. In reality, both planets have solid surfaces. Jupiter has atmospheric zones for thousands of miles above its surface, which balance and blend with the sun's energy to make a habitable environment. The earth—like surface of Saturn lies many thousands of miles below its atmospheric firmament. Many advancements in understanding of electromagnetic fields and cosmic science will be required in order to replace the old theories with correct knowledge.

George Adamski gave a complete explanation for the habitability on the more distant planets, and the following is taken from his book, Flying Saucers Farewell.

"One of the most frequent problems encountered when giving a lecture on space is the insistence of scientists that the outer planets are devoid of light and heat. Their objection is that radiation from the sun is so weak at these vast distances that Pluto, for instance, would be at absolute zero or close to it, with a frozen atmosphere, and totally incapable of supporting life— forms of any type.

This is the main argument brought against me when doubt is expressed about my meeting human beings from some of these other planets,

The first thing to realize is that the sun does not emit light and heat in the form we observe here on Earth. Radiation from the sun does not manifest itself as light and heat until it penetrates the atmospheres of the planets themselves. Outer space is devoid of light as we know it. The light in outer space is a cold light caused by the phosphorescence of vast clouds of particles and gases responding to the radiation given off by the sun. To a human observer, outer space looks like a dark, vast void, filled with billions upon billions of tiny specks of multicolored light. All of these tiny lights are in a state of continuous motion and activity.

Radiation from the sun is composed of ultraviolet light, hard and soft X—rays, cosmic and gamma rays. The greater portion of these destructive rays are filtered out by a planet's ionosphere and upper atmosphere. The innumerable, infinitesimal particles within a planet's atmosphere emit visible light when excited by the sun's filtered radiation. The earth absorbs these rays, and in return gives off infrared energy. Energy thus given off activates the atmosphere immediately surrounding the planet, thereby generating heat that keeps the planet warm.

It is easy to see how this energy from the sun can encompass our earth, after all, we are only 93 million miles from the sun. But what about the planets that are more remote from the sun?

What, then, is the answer? I know from personal experience that these outer planets have thriving civilizations, with climates and atmospheres similar to our own earth. The larger planets, such as Saturn and Jupiter, have much lower gravity than has been assumed by our scientists. Therefore, our explanation of gravity must be in error in some way.

Our main problem now is not gravity but climate. How do these planets receive enough energy from the sun to exist in a similar state to Earth?

A clue to this answer is found in the vacuum tube; more specifically it is found in the cathode ray tube. This tube, abbreviated CRT, is found in the ordinary home television receiver. In it we have a heater that raises the temperature of a cathode to a point at which it gives off electrons in great quantities. These electrons are negative in nature. High positive voltages are supplied to various grids and anode: in the tube.

There are 2 types of electricity: positive and negative. The electron is negative and its counterpart, the proton, is positive. Just as the north pole of a magnet will attract the south pole of another magnet, the electrons attract the protons.
Similar poles of magnets repel each other and so do similar charges of electricity. Likes repel; unlikes attract.

The high positive voltages on the grids and anodes of the CRT attract the electrons from the cathode. The electrons are pulled toward the anodes with great speed, but, due to the type of construction of these anodes, most electrons rush right on through toward the next one, Theoretically, this could be continued for great distances by use of several different anodes and high positive voltages.

Mercury, Venus, Earth and Mars are close enough to the sun to get good radiation. With the planets beyond Mars it is a different situation. At these distances the sun's radiation has started to diminish. At this time, it comes under the influence of the tremendous attracting force generated by the first asteroid belt which totally envelopes the central position of our solar system. The negative charge of the asteroid belt is great enough to attract the particles from the sun and pull them back up to their original speed. Because this belt is grid—like in construction, with thousands of openings and paths, similar to a window screen with air going through, the particles dash on through and enter the influence of the planets beyond.

These, being negative in themselves, as are all planets, attract from space the positive particles they need for light and heat. At the same time infinite numbers of similar particles rush on past and are attracted by a second asteroid belt between Neptune and Pluto, where the process is repeated all over again. This furnishes Pluto and the last 3 planets with normal light and heat. 12 planets in all exist in our system, according to the space travelers.

A third asteroid belt is beyond the twelfth planet, serving a dual purpose of blending space within our system with that of neighboring systems. At the same time, it serves as a protective filter, comparable to the ionosphere encompassing a planet.

We might summarize by saying: The 2 inner asteroid belts gather rays from the sun and speed them on through space. They equalize, so to speak, conditions within the system from the area of Mercury right on to the outer reaches of our solar system, while this keeps our system, as a unit, in balance with those beyond. Because of this cosmic activity, of which we on Earth have not previously been aware, we could go to any of our planets and enjoy a climate and atmosphere similar to our own.

Following this discussion, Adamski explains how the asteroid belt acts as a dielectric (as in a capacitor which stores electrical energy) by trapping a portion of the sun's energy while at the same time imparting energy to particles passing through. This alternating— current action effectively makes the asteroid belt an incubator of electro— magnetic energy. This net attraction of the sun's positively charged particles should register as higher kinetic energy (temperature) by a space probe passing through the zone.

Voyager 1 passed this zone that lies between Mars and Jupiter in 1978, and found it to be the hottest spot ever measured in the solar system.

The temperature was at least 100 times that of the sun's surface.

There is no real heat though, since the interplanetary region is a near —vacuum, whereas the plasma fields around the sun are extremely dense. Therefore, the Voyager probe was measuring high kinetic activity of solar particles in the asteroid belt, which is exactly the description given 17 years earlier by Adamski in his book.

The spaceships that Adamski went on were motherships from Venus and Saturn. The people he met on these craft were from Venus and Saturn, along with people from Mars the 3 planets most responsible for helping the Earth in its transition towards the space age. But he was told that there was interplanetary cooperation throughout our solar system, and that people of all planets except Earth travel space freely.

Every planet except Earth travel space freely. Every planet except Earth has spaceships.

Although Adamski's main contacts were from the same group, occasionally he met others from our system. While in Mexico in 1957, he had visits by space people who lived and worked in that nation. 2 that he met came from different planets, namely Jupiter and Neptune, Adamski related this information in a letter to a European correspondent, and stated simply that having met people from 5 different planets of our system, he found no more difference between them, than between various nationalities on Earth. Apparently, the information Adamski received was of a universal nature relating to their work on Earth. Adamski had explained in his 3 published books that the other planets besides Venus, Mars, and Saturn were Jupiter and Neptune.

Adamski explained that there were 3 undiscovered planets beyond the orbit of Pluto, making a total of 12 planets in our solar system. This information now needs to be revised to 13 planets, as planet 'K laden needs to be included here. This had first been revealed to him during a discussion on board a Venusian carrier ship, and stated in his book 'Inside the Spaceships' (1955). The simplicity of this statement lay dormant for 28 years. In January 1983, NASA placed a space telescope, the Infrared Astronomical Satellite (IRAS), in an orbit 560 miles above the Earth. For the first time ever, it revealed the outer reaches of our solar system, which had previously been invisible to astronomers because of the deep layer of atmosphere around our planet. High above the atmospheric interference, this 22.4inch telescope was so sensitive that it could have detected a 20—watt light bulb on Pluto. For the first few months, astronomers at JPL tested their computers, and by summer, released some sketchy details on the discovery of new comets, asteroids, and mysterious bands of dust circling the sun between Mars and Jupiter. Then the press was given a stunning disclosure in October.

 The orbiting telescope IRAS may have sighted at least one new planet, and more planets may come to light next week when astronomers gather here'.

There is a very good chance that IRAS will identify a new planet or 2 circling the sun beyond the orbits of Neptune and Pluto. 't said Nick Gautier, an infrared astronomer from the University of Arizona, -USA

Today, 10-31-83 The space agency censored so much information and since 1983 has been known as 'NEVER A STRAIGHT ANSWER' NASA disavowed any previous "un— authorized" releases, and shifted attention to significant findings way out in the galaxy. Although admitting that scientists had analyzed only 2% of the data delivered by IRAS so far, NASA only emphasized the "steady stream of wonders" beyond our solar system wispy clouds of interstellar dust and gas, the total energy in the universe, the un— known chemical soup from which stars are born, and the temperature of various galaxies.

The project director, Nancy Boggess stated, "A lot of chapters in the astronomy books will be rewritten when all the results are in. " but she added that it could take another 30 years to plow through all the information received from the year—long IRAS mission but the astronomy books are never rewritten to tell the truth about our own solar system. Space discoveries are censored by the agency 's top officials. In the case of IRAS, the Kitt Peak Observatory astronomers innocently disclosed the findings of planets beyond Pluto, not knowing the relative significance of the data. NASA of course, did know the significance, and therefore avoided refreshing any reporter 's memory at the press conference. NASA had conveniently diverted all attention to dull speculation concerning interstellar dust clouds and star systems that are light—years away.

What is the significance of the discovery by IRAS that was noted by the Arizona astronomers? It confirms the knowledge brought forward by Adamski in 1955. Adamski 's relationship to the UFO field is known to be authentic, and his space information regarding our solar system is known to be solid information. He was the only contactee who gave valid and specific information that could be confirmed later be it Venus, Mars, the Moon, or beyond. But for the space agency to confirm anything publicly, would also confirm the truth behind UFOs and our inter— planetary space visitors. This they have been instructed not to do, since certain powerful interests are behind the suppression of UFO information and 'man— made' flying saucers having been developed at areas S— 4 and area 51 in the Nevada desert...

NASA is also secretly following Adamski's guidelines, for the authorities know that they 're the only truths known regarding our solar system.
The criminal 'intelligence' agencies of the U.S.
government have prepared them with that information.

 For example, in 1960 a Pioneer probe on its journey toward the Sun registered 'magnetic' storms and disturbances 3 million miles from Earth, and found a gigantic magnetic field with an axis unrelated to the sun. This discovery by the experimental probe closely matched statements made by Adamski in 1955 in 'Inside the Spaceships'. In 1976 the structure of the sun's magnetic field was determined for the first time from data returned by the Pioneer 11 spacecraft, which was thrown 100 million miles above the ecliptic plane of the solar system after passing Jupiter. The probe revealed that the sun's magnetic field envelops the entire solar system in an elliptical pattern that is split into northern and southern hemispheres by a thin sheet of electric current. 15 years earlier, George Adamski related this description in his book 'Flying Saucers Farewell, and he further explained that these elliptical magnetic fields between planets could be likened to alternating current. These alternating elliptical fields, extending from sun to planet, are the invisible bonds which balance the solar system. There is much more that could be said on this intricate, balanced relationship that pervades and encompasses the whole solar system. To put it simply, a disruption in one part of the system will affect the rest of the solar system. A major disruption in the magnetic rivers" between neighboring planets could easily be caused by unrestrained nuclear power plants buildup and toxic waste pollution, along with nuclear warfare on Earth, thereby greatly unbalancing the cosmic fields of space. Not only would space travel by the space people be endangered, but their own planetary environment could be seriously affected. Now perhaps, we can better understand all the UFO activity by Venus and Mars in observing our Earth "come of age.

It is not generally realized that Adamski published, lectured, and conveyed an extensive amount of information beyond his 3 well—known books. In this additional material there is many a key for the researcher. It also shows a continual flow of new information from his contacts in the 1960's, as he developed a space education program alone philosophical and technological lines. Society's real growth can only be developed from true space information and new science applications. In UFO10gy, James McCampbell reviewed a case study of an encounter between an Air Force plane and a UFO. The original investigation of this incident was conducted by James McDonald and presented at the December

1969 meeting of the American Association for the Advancement of Science, with McDonald's analysis later being published in a book.

Although neither man knew it, their detailed discussions of this incident supplied substantial corroboration of information given by Adamski years earlier, regarding the propulsion of spaceships.

An Airforce B—47 was on a training mission using the military's top

ECM (Electronic Countermeasures) equipment, during a flight over the South— Central states in July, 1957, While flying at 35,000 feet, the aircraft commander and the co— pilot visually spotted a UFO, shortly after their navigation radar and 2 separate ECM receivers had electronically detected a rapidly moving object in the vicinity. The UFO suddenly appeared as an intense bluish—white light as it approached the aircraft. Thinking he might have to take evasive action, the flight commander was startled to see the UFO almost instantaneously change direction "and flash across their flight path from port to starboard at an angular velocity he had never seen matched in all of his 20-years of flying."

The pilot then received permission to ignore his flight plan and chase the object. He increased the B—47's speed to Mach 0.83 before turning to chase, and immediately the UFO pulled ahead. For the next hour, while covering about 800 miles, the UFO quite literally flew circles around the B— 47, by performing extreme maneuvers to show that it could easily outfly and outdistance the military aircraft. During this whole time, the pilot's visual observations of the UFO from the cockpit co— incided exactly with the UFO directions and signals deeeceed by the on—board radar and the 2 ECM receivers, and also by ground control radar from Texas to Oklahoma.

Readings from the ECM equipment were written as part of the official report of the UFO incident as follows. " • • •intercepted at approximately Meridian, Mississippi, a signal with the following characteristics; frequency 2995 MC to 3000 MC; pulse width of 2.0 microseconds; pulse repetition frequency of 600 cps; sweep rate of 4 rpm; vertical polarity... Signal moved rapidly up the D/ F scope indicating a rapidly moving signal source; i.e. an airborne source. . . James McCampbell summarized the above data into a single statement and said,

"This UFO was, in fact, pouring forth large amounts of electromagnetic radiation in a very narrow range of the microwave region and it was pulsed at a low, audio rate."

The following is taken from a transcript of a lecture and discussion given by George Adamski on May 4, 1955. There was a question from the audience asking about the frequencies involved in the running of the spaceships. Giving a general description in layman's terms, Adamski answered:

"The ships themselves carry frequencies from sixty to 70, 000 megacycles in full operation. Static electricity is electrical energy, like a speed that is not in motion. It is in motion, but not what we call motion because it is stillness. Magnetism is already propelling energy.

You have to convert static electricity into propulsion, or a pulsating state, in order to use it."

Note that the ECM equipment on board the B— 47 measured the UFO's radiating frequency at 3000 MC, which is consistent with the fact that the UFO was simply keeping pace with the 500 mile—per—hour speed of the aircraft. While in full operation, (and they have been tracked on radar flying in excess of 8000 mph in our atmosphere), Adamski says the frequency of the ship's energy would be 60,000 to 70, 000 MC. And his description of converting to "a pulsating state," agrees exactly with the ECM receiver detection that the UFO's electromagnetic energy was "pulsed at a low, audio rate."

The conclusions are simple. The highly classified incident did not come to light until
McDonald's report in 1969. The case involved the first direct measurement of a UFO's energy field, and it
was detected by sophisticated instrumentation on board a military aircraft. Adamski 's information preceded
McDonald's research, and
McCampbe11's notable review of the incident by
14 years. As Edward Ruppelt, the man in charge of

Air
Force investigations, once wrote, "What constitutes proof

? "

The pilot revealed that as soon as he landed following the UFO encounter, AF Intelligence went on board
the plane and removed all the electronic countermeasures data and radar scope pictures that had been
recorded during the flight. This was followed by intensive interrogation of the crew at the base
headquarters by the Intelligence personnel. Lengthy questionnaires were completed, and submitted as
part of the intelligence reports. The event and the information was classified in security beyond
classification, and the crew was told nothing further regarding the incident. The direct confirmation of the
energy associated with UFOs is always kept under the blanket of "national security."

With excursions into space by artificial means, our astronauts have to contend with the experience of
weightlessness. Pictures televised back from orbit always show the crew and unsecured objects floating
around the cabin. In its harmless banter, media coverage never fails to comment on this accepted, and
somewhat amusing, side— effect of "escaping gravity."

But in a related way, what is not laughable to the UFO censors is the description of true space flight given by
George Adamski in 'Inside the Spaceships'. Adamski and his space friends were not floating around the
spacecraft in a weightless condition, because the ships utilize the natural energy forces in providing a
gravitational field for the ship. And for the authorities to admit that the UFOs in our skies are interplanetary
spaceships, would be to admit that this type of energy is within our immediate grasp.

These spaceships are utilizing the natural electromagnetic energy of space for propulsion and for over—
coming gravity. We could readily harness this natural energy for all our domestic use, including land
transportation, heat and electricity. It would be as free as the air we breathe. Official acceptance of the
authentic reality of UFOs would in turn lead to the acknowledgement of their free energy, and this would
break up the economic system of the money changers.
So they fight it.

The world's energy controllers have always been the major force in blocking the truth about UFOs from
coming out. This cartel is made up of the oil corporations, the power companies, and especially the
criminally insane nuclear energy industries. And this cartel is really above and controls the international

bankers. With their complete dominance of the world's energy resources, the ongoing stability of the world economy is subject to their control, and this forces the many governments, both capitalist and communist, to do their bidding. Those that fear the truth coming out are those in control of the monetary system. Because if the propulsion of these spaceships became known, the energy controllers would be ruined, and their major polluting industries obsolete.

The war profiteers are also a major force behind the cover-up. The world's military spending now totals many trillions in Euro$ and the bankrupt U.S. economy, with the United States and Russia counting for more than half the buildup. The superpowers keep fanning the flames of war around the globe, because international marketing of armaments and nuclear power stations and weapons are the basis of the world economy. The U.S. national debt is in the trillions+++++++ and this debt is placed on the people, while defense corporations are making billions. There can never be real peace with a system of distorted wealth, be— cause peaceful co—existence would disintegrate the privileged power system of the money changers. The world's population is constantly kept hostage under poised nuclear weapons, in order to serve a few money changers.

Life is as close as the Moon and planets, but our world will be on the brink of Armageddon and never know that the life is there. The slightest admission that.
there are peaceful civilizations bringing their visiting spaceships here to our planet would be devastating to the money system of the war profiteers.

So together, the energy cartel and the weapons industries have every— thing to lose, and therefore they do not want the truth to be known. The truth of UFOs and the truth of space would entirely expose how wrong cur global priorities are.
These powerful corporations are the force behind the international Silence (criminal) Group, which has suppressed the truth about UFOs and censored the real knowledge behind planetary findings ever since the 1950 's. This Silence Group has an interlocking chain of invisible control working through government intelligence agencies, mufti national corporations, and even national institutions.

This formidable group is in essence, a secret international government that serves the interests of the energy controllers, weapons producers, and bankers.

They have many ways of shaping events and many areas of influence to control. In their powerful conspiracy of world economics, they have many things to contend with, but we will restrict this discussion only to the issue of UFOs. The cartel has its own clandestine agencies working within the intelligence organizations of world governments.

They have used every means to control and conceal the real UFO evidence from the public and scientific bodies. At the same time, this Silence Group insures that regular publicity is given to weird tales and crackpot ideas, thereby keeping the idea of UFOs in a ridiculous light. They even plant obvious hoaxes in the press, and foster false contact claims through the media by paying imposters. The opposition wants the public to associate UFOs with cults, and with psychics claiming to receive bizarre messages from space "entities." Any crack— pot scheme is played up to instill a sour impression in the minds of the public.

But the biggest part of the censorship extended into the national space program.

Life and living planetary conditions beyond the Earth, which could have been revealed from the earliest probes, was totally suppressed. A lifeless, uninhabitable picture was deceptively documented, and public thinking manipulated, in order to protect the money system of the world. Left in ignorance about space, the masses have been unconsciously led to electing dark leaders who teach hate and revenge, who build more destructive weapons, and who aggressively lay the groundwork for wars and threats of war. The early adventure of space has been replaced with the present militarization of space. We are now marking time. We are closer to biblical prophecy than the other road which could pave the way towards a glorious future.

What is the purpose behind this secret international government? It is to keep the monetary system from changing, despite whatever progress mankind makes socially.

The masses will not change their thinking, unless there is a change coming from institutional thinking, or direct information on space from government bodies.

Nuclear weapons, nuclear power plants and Star Wars development are promoted today — only to defend our system of money which is a false god.

Political leaders and their clever sophists have fully desensitized the common people to the constant talk of wars and arms buildup. Any pretext or propaganda used to fan fires and identify "enemies" is acceptable. The masses have been conditioned to go about their trivial pursuits and directionless lives, and to leave their collective fate to the gambling men of war. Perverse power has been gained by the sudden exponential increase in arms selling up until now by the super— Q4q (weapons on demand) powers.

We are Atlantis today, slaves to habits from the past, and worshipping military might. Our fate is being decided by the false prophets of war and commercialism, and the dark forces in political power wield control by keeping the minds of the public confused, uncertain and in fear.

We know that the space shuttle is for military purposes. It has nothing to do with space exploration or learning, but instead serves as a means to bring our war technology into orbit around the Earth it will sometimes carry incidental scientific cargo, as part of the public relations hoopla, but the shuttle has really been developed for militarizing space around the Earth. It is not designed to get man into space, but to get man's weapons into space. Therefore, it will not contribute to the advancement of man.

If the authorities were to admit the truth about UFOs and space, our whole approach would be forced to change. And rockets would now be outdated. In fact, there is no reason, except for economic ones, that we are still using rockets to artificially overcome gravity. We will never advance in capability, and we will never become a space civilization long as we are stuck with rocket launching. It is primitive technology, and it opposes everything that this government knows about UFOs and their technology, and the government knows it. So do top scientists and the military.

Rockets are anything but true space machines. They go in one direction, one time, with a tiny crew strapped to tremendous G forces. And they are such an artificial way to overcome gravity, that the tremendous cost is a devastating expenditure.

Often our space visitors have been seen flying in huge formations of ships; occasionally a hundred or more would fly around an area together. Can you imagine seeing 100 U.S. space shuttles flying together over California? The cost would be a trillion dollars. You would not see them hovering, nor would you see them for very long. They would soon be running out of fuel, be forced to land, then have to be towed to a launching facility. Each would have to be hooked up again to multi—million—dollar booster rockets and external fuel tanks, and 100 earth—shattering blast— offs would be required to put them temporarily back in the sky.

This chapter is perhaps written a bit forcefully, but it is time to shake things up a little. We are out to counteract the insidious cover-up regarding space conditions that has been perpetrated on the masses, leaving society lost and hopeless. "When goodness is mocked and good people thrown into despair; when the false, the cruel, and the diabolic are so powerful, it is not surprising that men's minds are stabbed with doubts. (Paul Brunton) Men can only recover their faith in humanity and in themselves when the truth is known and understood."

Back when the UFO situation was a new and challenging field, there was seemingly a luxury of time to ponder over the information, fifty— six years have changed all that. Our present world situation is acutely critical — precariously balanced on the edge of a cliff and its imperative that as much information as possible be brought out to the public regarding the truth behind the UFO evidence.

One should readily understand, that the Silence Group does not care if a few people learn the truth about planetary science (Venus, Mars, and the Moon), or if 1,000 people learn, or even perhaps 10,000. But if the common people learn the truth — if they are told then there is a problem for the censors, because the number could soon reach millions. And the energy controllers and the billion— dollar defense industries would no longer have the unquestioned mass support that they always have had.

This is why Adamski 's evidence and space information posed such a threat to those in opposition to the truth. His information reached millions of people, and society was waking up to the beauty and splendor of life and space. Adamski always advocated the benefits of our becoming a true space civilization. Mankind would advance in learning and understanding unlike any previous time in history. Our planet is just as important as any other planet, and it is an integral part of the household of life. Any dream one may have, must have bene— fits here on Earth. Not somewhere else.

Adamski always emphasized the need for our civilization to shift to a realistic space economy, and have interplanetary spaceships built here on Earth. Then man would be learning all the time, and becoming a true space civilization would lead us to greater cooperation and harmony on Earth. This would steer mankind away from his nuclear experimentation against Nature, and save this civilization. Knowing the truth about space sciences and the truth behind the UFOs, were the things that would advance man's understanding in order to make the Earth a better place to live.

That has always been the importance of this field.

It is man's immature idea that the UFOs are an invitation for us to get out in space and we'll just leave the mess on Earth behind. No. It means work with each other and work with Nature here, in peace, so that we will be able to join in the adventure of space. The Earth is our home, and it is time we treat it with respect, and treat our brothers and sisters in every country with respect. Then we will be welcomed and helped to get in space. True space travel and visits to other planets would be our way of life.

The planets in our system are inhabited, and people of all planets except the Earth travel space freely. With their ships utilizing Nature's magnetic fields, the distance between planets can be covered within a few hours to a few days. Most spacecraft seen in our skies are from Venus. Smaller numbers are from Mars and Saturn, followed by visiting craft from other planets in our system. The space visitors are peaceful and friendly, and certainly human, but also very protective of us because we are all one interplanetary family in this solar system. But we on Earth do have to start earning our way a little, because open friendship cannot be extended if we refuse to believe in their existence, and as long as we do not express the will to live peacefully among nations on our own planet.

There is a chasm between knowledge and ignorance which the arches of science can never span. Thoreau
Z52 (Constant Alert)

CONCLUSIONS ABOUT OUR INHABITED SOLAR SYSTEM

Many insights into the remarkable character and lifetime work of George Adamski have been revealed through writings and discussions by his former co—workers and associates. To review all the information which I have read, and that which I personally learned through contact with several of his co—workers, would fill the contents of another book. Adamski also had many meetings with high government and space officials, and while in Europe he was received by Queen Juliana and Prince Bernhard of the Netherlands. In 1963 Pope John XX III gave him a special golden medallion, that was only given to a very few people for high humanitarian achievement.

Perhaps the following example of mine will reveal a little about the nature of the man a man who had set out on an extraordinary mission

to speak on behalf of the space visitors coming our way. It was no easy task in the 1950's, with all the prevailing skepticism, the organized opposition, and the general insistence of scientists that life on neighboring planets was impossible. But Adamski faced all the trials and tribulations of such a mission in a dignified and respectable manner, to give the message, and to give expression to the truth by living it.

The example I wish to cite occurred shortly after the publication of 'Flying Saucers Have Landed'. The book itself was unusual in that its 2 authors had not met in person until long after initial publication and several reprints of the successful volume. (Desmond Leslie and Adamski had agreed by mail to Adamski 's manuscript and share co— authorship). The event took place in the summer of 1954, when Leslie made a trip to
Palomar Gardens, California, to first meet
Adamski, his co—author

One evening, while Leslie and a few others were having a quiet d is— cuss ion on Adamski's patio just after dark, Leslie recalls that all of a sudden he got a tremendous feeling of being watched. He swung around in time to see a small golden disk not more than fifty feet away, which then shot upward in a trail of light. Leslie was amazed to see a remote scanning disk from a spaceship, yet he relates that Adamski did not alert him to its presence although he knew that it was there all along. Only after the disk had left, did Adamski grin solemnly and say ' "I was wondering when you were going to notice that."

Whereas someone else — a lesser character involved in the controversial field of space visitations would have undoubtedly exclaimed upon noticing the scanning disk: "See that! I want you to see that, which will prove me right, (etc.). " George Adamski never worried about so— called proof. Regarding his space contacts, the proof was in the message and knowledge given by the space people. Adamski gave express ion of the truth, and presented everything on a practical and understandable level for all people. Following the publication of 'Inside the Space
Ships', Adamski was able to develop an international Get Acquainted Program that could bring space science information and space age philosophy to citizens in many countries.

If one looks at the complete record, it becomes apparent that George Adamski was the most qualified person ever, to be contacted. And in retrospect, it can be presently understood that the space visitors did give the full information behind their coming. It was given once. True, there were progressive developments over several years, during the period we began to enter the space age. And the hundreds of thousands of UFO sightings (existent right up to the present) continued to alert the public to the reality of space visitations, and to support that information. But the truth behind the space visitations was given in its entirety once — it will not be repeated. In other words, there will be no new answers to the identity or origin of the spacecraft in our skies. The majority are from Venus, and the other spacecraft that are cooperating in their interplanetary ventures are also originating from planets in this solar system.

This was quite acceptable in the early years, but became more controversial as official announcements regarding space conditions mounted up during the 1960's and 70's. The negative image of planetary conditions was based on dogmatic perceptions in orthodox science, that in turn was supported by official announcements regarding space probe missions that were heavily censored. Has officialdom gone beyond the point where it could be straightened out by objective and value— free science?

George Adamski presented the truth of our Venusian space visitors, and he provided actual descriptions of the planetary environment. Soon after his initial contact in 1952, he had been allowed the privilege of a few trips into space aboard their craft, to learn what he could and then give the information to people on Earth. While on a mother— ship on August 23, 1954, his space friends showed him live scenes of Venus by using a projection system so advanced, Adamski could scarcely describe how it was done. And the pictures and scenes of Venus were vivid and distinct, colorful and dimensional.

Adamski saw magnificent mountains, some rocky and some with snow. Many were thickly timbered, with streams and water cascades running down the mountainsides. He was told that Venus has many lakes and seven oceans, all of which are connected by waterways, both natural and artificial.

They showed him several Venusian cities, that all followed a circular or oval pattern. People on the streets were going about in the same manner as we do, "except for the absence of rush and worry so notice— able with us. " Their clothing was similar to our own. There were conveyances for mass transportation that silently glided down streets bordered with beautiful flowers. Adamski was shown beaches and tropical areas with bird and animal life. He also was informed that people rarely see the stars as we on Earth do, because of the firmament of the Venusian atmosphere.

These descriptions were contained in his book, Inside the Space Ships'. In a later account, he mentioned that a day on Venus was 26 hours long, and that the atmospheric pressure was identical to that on the Earth.

We know that the Russian and American space programs sent a total of a dozen probes to the planet's surface, Common sense should tell us how the situation was handled on their end (by the Venusian civilization). But perhaps a short discussion given by Adamski in a private report will give a complete understanding to us. The information was included in a special letter by George Adamski to his close co— workers in the spring of 1961. At that time, the very first interplanetary space probe had been launched from Earth and was enroute to Venus.

The probe was Russia's Venera Spacecraft, and it had been launched on February 1 2,

1961. (Earlier probes by the U.S. and Russia had all been directed towards the Moon).

Adamski 's discussion was based on both public press reports and in for— mat ion obtained from meetings with space people at the time. He wrote: Russia now has a space probe enroute to Venus, although we have been informed that she has lost contact with it. It is for scientific purposes, and since this is the only way we have of securing the knowledge we need, it will be allowed to land on that planet if it is accurately directed... I was told that a large interplanetary ship is following the probe sent out by the Russians, but not interfering with it. Much has already been learned from it (the probe) by our scientists. For instance, the date on which this space probe was expected to hit Venus was at first set for some time toward the end of May. Then it was set up to early May, then to mid— April. Later, one of our commentators reported it is nearing Venus. The reason given for these changing dates was that the probe was caught in some unknown force from the sun that was pulling it in faster than anticipated.

This part was verified by my space friends, and pointed out as only one of the infinite number of factors with which we must become familiar regarding outer space to be able to travel it safely. They did not say whether or not this sun's force had drawn the probe sufficiently far off its original trajectory towards Venus. Logically, this could happen. Because we must learn from our own experiences and from our own instruments, the space visitors, while observing and re— cording the results of our efforts, are not interfering nor passing on to us their information. This is as it should be, because lessons learned from experience are better understood and remembered, as we all know.

If the Russian calculations are accurate enough to enable their probe to enter the pull of Venus, it will be allowed to land there. Should its course be in direct line with a city or community of any kind, it will be slightly diverted to hit in an uninhabited part of the planet, thus protecting life and property. We would do the same if we were in their position. The diversion will be so slight that it will not interfere with the recordings sent to the probe's home base.

Let us say that this space probe lands safely in a desert area — and there are some on Venus similar to those on Earth — its instruments might conceivably send back reports of "no life". We neither know what instruments are enclosed, nor how they might survive such a landing. I strongly suspect there is no camera included that could photograph and send back reports of a person or many persons, who would naturally go to investigate this strange object. You may rest assured that it will be examined, however, with as much interest as we would give to an object under similar conditions. Perhaps even more, because they will have been alerted as to its nature even before it hits.

On the other hand, let us assume there is a camera within the probe that can and does send back photographs of humans, animal life, and various types of vegetation. Such information will not be made world— wide knowledge immediately. There will be reports and contradictions for a long time to come, regardless what nation is successful in reaching another planet first. would go so far as to say that even when men and women are sent and safely land on another planet, or on our moon, they will be instructed to report and show only so much, should they be hooked up with a world radio or TV chain for reporting. Other information will be noted and given in personal reports to their government on their safe return.

This will probably be due to both the religious and political conditions existing throughout the world, and a fear as to how much could be accepted by the multitudes. Although I believe it may be only a relatively short time before we will have men traveling space in ships of our own making, T doubt that we will soon be given the whole truth of conditions found. So do not be disturbed by conflicting reports, which will be increasing in number as time passes. Remember how important events have been handled in the past (and take all reports in stride). Time will prove, that information given to me and which T have shared with you is true fact.

I had thought naturally along these lines for many years, and only recently did I receive a copy of the complete letter from which this report was taken. With regard to Venus probes, the report provided a logical explanation as to how that planetary civilization would respond to our efforts with remote space vehicles. The situation would not have been any different with probes to Mars and the Moon the only two other planetary bodies that have been probed by landers.

The planetary civilizations would not interfere with our simple devices sent out to check up on surface conditions. Mankind is allowed to follow his own chosen developments, and the interpretation of whatever he can detect in his reach to explore the unknown spaces beyond the Earth. I have given in this book just a brief summary of actual evidence behind space science research regarding the near planets, and the resulting limitations of our present—day knowledge due to several reasons.

Acknowledgement that there are human civilizations on our neighboring planets, though profound to realize initially, is by itself not going to teach us anything about those space civilizations, or how we are going to achieve that society growth in our own world. It is only a first step. Scientific technology also, is but a small part towards becoming an advanced society. The mere fact that we can use rockets to put a satellite or a military payload into orbit does not make us a space civilization. The mentality of the world must change first. Right thinking must be developed towards a reverence for all life. We only give lip service to the ideas of peace and brotherhood, all the while supporting the very things that prevent this from being a reality. We will not become a true space civilization until we respect each other as fellow human beings, recognizing all as equals, brothers and sisters of one planetary home.

Then, properly guided by humanistic thinking, people would actively support true space knowledge, that would advance our civilization and begin to establish a peaceful, productive society on Earth. The real mysteries of life would be replaced by the reality of life the reality which is understood and lived by the space people coming our way. Mankind can presently be alerted to the truth about our neighboring space civilizations, and when our society begins to develop purposefully along these lines, we will be able to go on and become the greatest and only enduring civilization this world has ever known. In conclusion, I really hope that this book has provided many details on the subject of NASA's hidden agenda in the militarization of space along with the subject of UFOs and space science for many people. The many pages about 'KARMA', with all of its fascinating aspects mentioned in this data should wake up many people to look within them— selves to realize their own potentials in life. It merely points out a direction to follow.

For personal understanding and individual growth, one must seek to learn self— knowledge, so that he or she may develop and express his or her inborn talents, and better serve their fellow— mankind. "Man know thyself" is the requisite for wisdom that has been handed down through the ages. With this understanding, and above all else in his recognized accomplishments, George Adamski brought out the reality and the truth of life, to the public.

As a society, and in reference to the whole field of flying saucers and the 'space sciences' and space visitations, we are presently at the same point Adamski established in the 1960's. He completed a program which fully presents an understanding of space age philosophy and science, which if applied by society, will go far in establishing a peaceful and advanced civilization on our planet. This program of space knowledge and cosmic science has been carried on by his close associates and former co—workers since 1965.

I would like very much to hear from the readers, as to their interest in the field of 'KARMA' and the other related science matters in my book. With your additional reading, one will soon learn the full importance behind the visitations of interplanetary spaceships, and the responsibility of our present—day civilization to bring about true peace. If enough people are informed of the truth regarding space and the space sciences, our society will then make a turn, and this planet

Earth will go on to become a true space civilization in our solar system

Prof. Dr. Hans J. Petermann can be reached at his home phone: (760) 327—4761 for any questions the readers may have pertaining to the data in his book. Please note the reference section for any additional information.   HITLER'S ASHES, UFOs, and SECRET WARFARE

Despite a popular perception of Antarctica as being a frozen waste— land, there exists a largely hidden history of a different Antarctica, of a habitable (thanks to geothermal soil heating) zone in Antarctica, of a Nazi— controlled piece of Antarctica where a heavily armed colony, it is said, still exists. Crazy as this may sound, there is evidence of at least Nazi involvement with the area, some as close as an older atlas. Curious? Break out an earlier atlas and take a look at the map of Antarctica. Look for the region called Queen Maud Land or
Dronning Maud Land. It's part of what used to be called Neu— schwabenland (New Swabia Land), a title given by the Germans under Captain Ritscher, who claimed the region for their own back in 1938—

1939 flying 2 Dornier Wal (Whale) float planes from the survey vessel
Schwabenland, and some atlases still carried this label in parentheses during the late 1970s and early 1980s. How? By meticulously photo— graphically surveying the region and dropping weighted marker posts topped by swastikas to delineate their claim (and magically capture it). Interestingly, Admiral Richard Byrd was invited along, but declined. Research for this article at www.south— p01e.com/p000019.htm found that he was busy preparing for the United States Antarctic Service Expedition 1939—41

According to numerous sites, many in German and read in translation,
Neuschwabenland became and was treated as part of the Third Reich. Moreover, it became the site of intense covert scientific research and military colonization, under the name Base 211, starting around 1942. The base was situated in the aforementioned ice—free area and reportedly also enjoyed underwater access from the sea via ancient volcanic tunnels, making it perfect for U— Boats. A map of this area is on the back cover of my earlier book entitled 'Gravity, Matter & Space Travel' in great details. You need to read this volume as a must to properly understand the importance of all of the underground earth cities worldwide. This author visited the inner earth cities that are mentioned, especially in Bolivia and in

the caves of Aggtelek in Hungary, crossing over into Slovakia. . . There is a unique photo of this cave in the book in glorious living colors with some unique features!

What's also remarkable about this area is that a range of sources explicitly link it to German flying saucers. These sources include: W. A. Harbånson's massively researched and footnoted novel 'Genesis , in—formation from German secret societies and Top Secret
SS blueprints in the video "UFO Secrets of WWII German Flying Saucers," excerpts from Rear Admiral Byrd's purported diaries (personally confirmed years ago to this writer by Byrd's nephew, Harley Byrd, and buttressed by statements from Al Bielek, who told me in 1993 that Hitler's saucer programs bankrupted Germany), and a series of shocking interviews given in 1947 to various reputable South American newspapers by then Rear Admiral Byrd, who warned of a "threat from the poles. Such candor came to a screeching halt when he got back to the States. What prompted such actions? If the various reports are credible, he lost a bunch of R4Ds (naval C—47s equipped for photographic survey, six were taken down on the aircraft carrier S.S. Philippine Sea as part of the

13—ship Operation High Jump http://www.usap.gov/videoclipsandmaps/mcmwebcam. cfm military expedition and flown to Little America, the U.S. Antarctic base) to hostile action and was shown by German—speaking saucer crews supposedly even in face— to—face meetings, in no uncertain terms, just who had the power. The hidden histories indicate that he didn't take this lying down, returning a decade later, under cover of an International Geophysical Year expedition, and delivered 2 nuclear weapons. on Base 211, wiping it out in the view of some and not wiping it out in the view of others. These actions are alleged to be the real reason for the hole in the ozone layer.

Research for this data also turned up a German book
'Flugscheiben Mber Neuschwabenland Flying Saucers Over Neuschwabenland. As if this weren't enough to consider, we must also take into account Grand

Admiral Karl Doenitz's (headed German government after Hitler 's suicide'?) claim that the Kriegsmarine (German navy) had built an impregnable fortress for the Fåhrer" in a faraway location and the statements made to author and WWII veteran Howard A. Buechner by a former SS officer who said that he was on the U—977 (known to have finally surrendered in Argentina months after the war ended, along with U—530) and that it went to Antarctica on a vital Nazi errand.

What errand? As related in Adolph Hitler and the   Secrets of the Holy

Lance (Thunderbird Press, 1988) and in 'Hitler's

Ashes Seeds of a New Reich
(Thunderbird Press, 1989), the cargo was indeed
Hitler's ashes, the notorious Blutfahne (Blood

Flag, as in Nazi "martyrs" Horst Wessel, etc.) of the Munich beerhall putsch (seen in the Nuremberg Rally footage being used by Hitler to magically pass its essence to and "bless" through the Law of Contagion)

the flags he touched with it), and the real Spear of Destiny AKA Lance of Christ. The one found by the Allies was reputedly an expert forgery by a Japanese master swordsmith.

In occult and practical political terms, he thus reportedly carried with him not only the core symbols of Nazi ism, but the ritual tool by which to resurrect it and give it real effect — the Spear of Destiny. Since the book does not mention a base, why use Antarctica as secure storage if there wasn't one? Even more bizarrely, if we accept that the Nazi core was stored in an ice cave, how was it found in the 1970s when the Blutfahne and the Spear of Destiny were supposedly slipped back into Europe after a clandestine light aircraft visit to Antarctica by the same former SS officer? This objection makes eminent sense considering Richard Byrd was barely able 'to locate Little America only a few years after leaving it because every— thing but part of the smokestacks and the radio tower had been buried by the snow. The former SS officer asserts that the Spear was to be used to transform Europe. Did this alleged operation have anything to do with the fall of the Soviet Union in 1989 or the rise of neo-Nazi movements in Europe? If the timeline's correct and a dummy was left for the Allies to find, then there was plenty of time to replace the dummy with the real one before the recent forensic investigations were carried out on the Spear, investigations which confirmed parts of it dated back to at least A.D. 1100, and the jury's still out on the age of the reputed nail from the True Cross said to be in it.

THE LAKE VOSTOK ANOMALY

In more recent times a powerful magnetic anomaly has been detected in the Lake Vostok region. Some believe the U.S. military personnel?

may have created a secret base there. There is also at least one volcano there whose eruption, it is suggested, could melt the Ant— arctic icecap. All these items have been reportedly confirmed to

Thomas Greanias by National Science Foundation personnel, as stated in 'The Secrets of Atlantis Revealed: Frequently Asked Questions http://www.mirror.co.uk/news/technologyscience/science/secrets-britains-atlantis-revealedarchaeologists-6361422

Richard Hoagland's Enterprise Mission http://www.enterprisemission.com/ noticed the apparent existence of some sort of medical crisis among U.S. personnel involved in the work. According to several articles there, the numbers of bizarre, urgent med evac cases among the care— fully screened staff suggest biological and/or radiation exposure. Weirdly, this is apparently backed up by a January 16, 2003 Pravda.com Russian story

http://www.pravdareport.com/main/2003/01/16/4214

[%204.html](#) titled "Scary Secrets of the Third Reich's Base in Antarctica, " where we learn "a research expedition discovered a virus in Antarctica" that neither "people nor animals had immunity to" and which could "cause a catastrophe" to the planet. Gulp! While some believe these unknown viruses are naturally occurring, others believe the Germans may've done covert biowarfare testing here, with the viruses as the engineered agents. Another possible scenario is arrival on a meteorite. In the meantime, several nations (Russia, U.S., U. K. FR. G., and Italy) are working on plans and activities to drill into mysterious Lake Vostok (Russians already down to last 328 feet (100 meters), with plans to halve that beginning soon and other Antarctic lakes, using environmentally sterile drilling and robotic exploration technology. Scientists consider that these sealed ecosystems, which probably contain life, may mirror conditions on Jupiter's satellite Europa.

Who can say what discoveries may lie ahead. If Atlantis awaits, could it be just the tip of the iceberg?

As mentioned previously, Ganymede definitely has been inhabited for a very long time and their UFOs have visited this planet many times! Please refer to my book "Gravity, Matter and Space Travel" for many more photographic details concerning our neighbors on Mars, the Moon and our other inhabited planets like Venus and Saturn.   ANTARCTICA REVISITED

The author Henry Stevens has dug very deep to uncover a much deeper mystery.

Geologically speaking, Antarctica is almost split in half by a huge rift valley, which runs from the South Pole toward Africa, then up East Africa to the Dead Sea. This entire rift valley is overflowing with geothermal activity. It was on this very rift valley that the Germans located their Antarctic base. Hot water ponds, named the Schumacher ponds by the Germans, are teaming with algae, and found on surface rock deep within Antarctica. These ponds never freeze over. Interestingly enough, each pond is populated by a different species of algae, giving each pond a different color. It is not far—fetched to believe that a sustainable base could be located over one of these geothermal vents, especially deep within a large crevice or cave... The Icelanders rely on geothermal energy to produce electricity for their daily needs. Why could not have inhabitants of Antarctica?

Adding to the strangeness, Stevens reports that an anonymous letter was sent to the journal 'Scientific American' in which a curious incident was alleged to have occurred in Antarctica. Two Australian women were attempting to cross— country—ski over Antarctica when they were seized by American Navy special forces, and detained.

The detention may have something to do, argues
Stevens, with all the strange secrecy that the United States shows towards Lake Vostokr a vast underground thermal lake in the Antarctic interior, covered with an ice dome that, during the endless day of the Antarctic summer, admits enough sunlight to bathe the lake in an endless twilight glow of light coming through the ice dome. But why was no less than the American National Security Agency involved? Stevens maintains that it is because it may be because Admiral Byrd and the Nazis may have found or discovered something there during their 1938 and 1947 fly—overs and photoreconnaissance of the continent.

This may very well be the truth, for there is a huge magnetic anomaly on the southwestern shore of Lake Vostok, registering some "1, 000 nanoteslas of variance with the surrounding vicinity. This might be due, as Stevens argues, to the presence of a vast amount of metal, "Metal as in a buried city. "

Or perhaps a genuine geophysical anomaly? Or a buried something else?" Or some combination of all three? We will never know.

But the evidence, Stevens argued, points to 'something' artificial, and something moreover under intelligent control. Consider, he says, the following Antarctic seismographs. These readings were taken from the American station at the Amundsen South Pole station, whose seismo— meters are located in such a fashion that they are "surrounding the old German area called Neuschwabenland. The first graph is that of a typical "all quiet on the Antarctic front" sort of day.

And here is what an Antarctic earthquake looks like:

But then there appears to be, according to
Stevens, highly anomalous, i.e. 'not naturally—caused, standing long—wave activity, when the needles are all over the chart in what is definitely 'not' an earth— quake, nor, for that matter, an explosion of any sort:

As if that were not enough, then there are signatures both of these standing long waves, and of small, sharp anomalies that would be typical of the seismic signatures of nuclear detonations:

PMSA.09LH2 Start Date:03/23/03 Filter: band-pass Displacement Magnification = 3000.30 @ 0.020 Hz

"All Quiet on the Antarctic Front "89

"Open War on the Antarctic Front": Standing Long

Wave Activity
Accompanied by

"Detonation "

Signatures.92

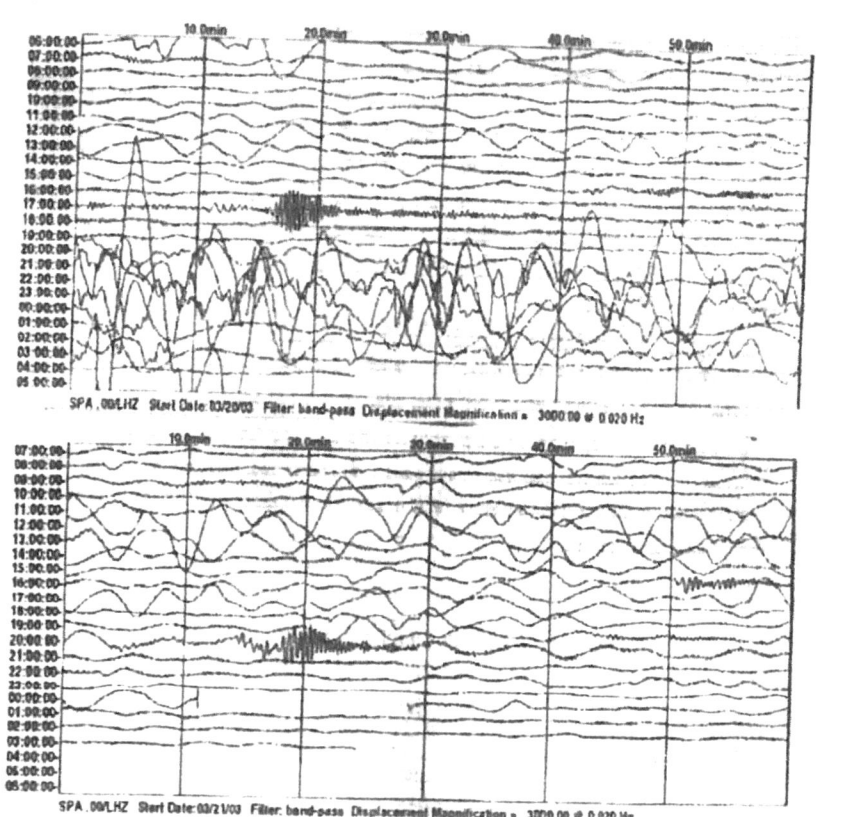

"Not So Quiet on the Antarctic Front":

Standing Long Waves, a Scalar Weapon Signature? 91 Typical Antarctic Earthquake Seismogram90

Stevens cites the research of German seismic researcher Christian Saal, who "interprets this to be an American attack on Neuschwaben— land using the new boring atomic weapon announced at the time of the Iraq

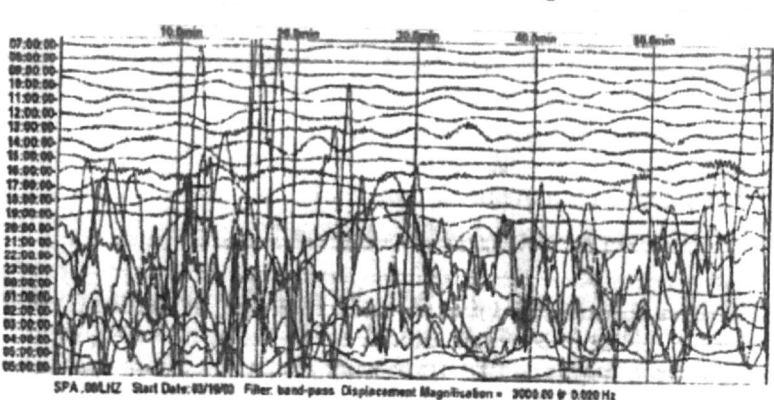

invasion."

Moreover, adding credence to this idea, the date on the first seismograph at the top of this page is 3/20/03, " the very day the United States began its massive bombardment of Baghdad." Saal maintains that while the world's attention was diverted to Iraq, the U.S.A. used "bunker— busting" atom bombs to attack the Nazi base. The defense, according to Saal, apparently held, because there was a second such "attack" launched, as is 't demonstrated" by the second seismogram. Stevens, of course, is much more

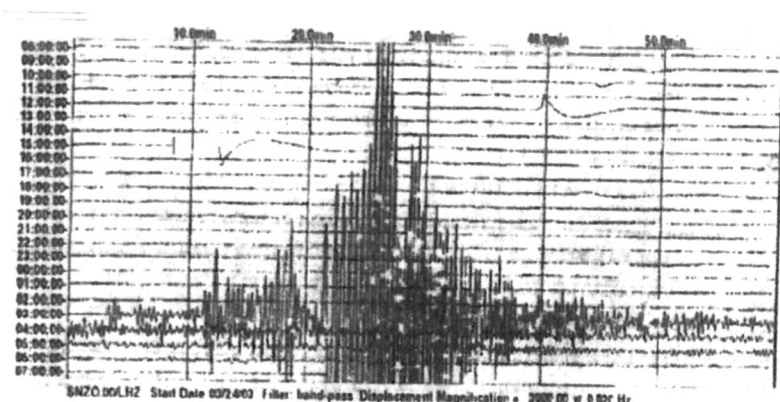

skeptical, though con— cedes the overall importance of the seismograms:

I really don't know if Christian Saal's interpretation is correct or not. What is most important to me is that there is real data here that shows something unusual and unexplained by "the authorities". This hard seismological data, an acknowledged mystery lake in the center of
Antarctica, and the interjection of the National Security Agency all point to something as yet unexplained happening on our southern—most continent.

Whether or not the current activity has anything whatsoever to do with Nazis or their nefarious schemes is, in a certain sense, not even important, for it seems hardly coincidental that they would have chosen to explore precisely this area, nor coincidental that they also researched the very technologies of standing long waves evidenced in the seismograms.

But whether these seismic anomalies point to secret buried cities, to a flying saucer long ago buried under the Antarctic ice, or to geophysical processes understood by but a few, or to some combination of any of these, one fact remains, and that is, within the wider con— text of Nazi saucer research and all the other exotic technologies that remained in their hands after the war, after all, there might be truth in the old mythologies of the Last Nazi Battalion, Antarctica, and secret saucer bases.

FLYING SAUCER MOTHERSHIP

This design is of a 'mother ship' with a diameter of about 250 meters. According to the scale at bottom right below see car & yacht the discus form is ideal for a maximum cooling system & minimal front surface, with a maximal content surface being a universal solution for interdimenensional cosmic speed travel.

The propulsion system shows a solar disc   2 particle accelerators spinning in opposite directions to create its own gravity field of weightless energy. The hydrogen molecules are converted into helium. This requires a small angle of incidence, using a strong & very flat pressed disc or triangular form. The 2 auxiliary plates are smaller landing units attached to the mother ship in this design. The great electro— magnetic fields & high temperatures of the vehicle's skin would make it very difficult for the people of the earth to see it clearly, due to the ionization of the surrounding air. There is a need for building 2 smaller discs first. The idea of this kind of space vehicle would be great to develop in comparison to the vehicles now being flown by the U.S. government, namely the V TR—3B l which has been a part of the super— secret U.S. Aurora program for many years!

Explanation of the numbers:

 Armored tank ring main propulsion drive system   tangential connections between main propulsion system & connecting ring.

Head braces & rings

Skin surface & cooling systems

Pressurized system

Room for devices

8. Fuel tanks (water) of the mothership

Crew compartments, eating facility food storage, etc.

Remotely controlled landing unit 'tank discus" for wet plänets diameter about 80 meters or smaller size(s).

1 1. Water tanks of the landing unit crew area of landing unit

movable command tower direction of constant travel to slow down or increase travel speed of mother ship. Comparative sizes of car and yacht G17 two evacuated rings in which aether is flowing super— conductors.

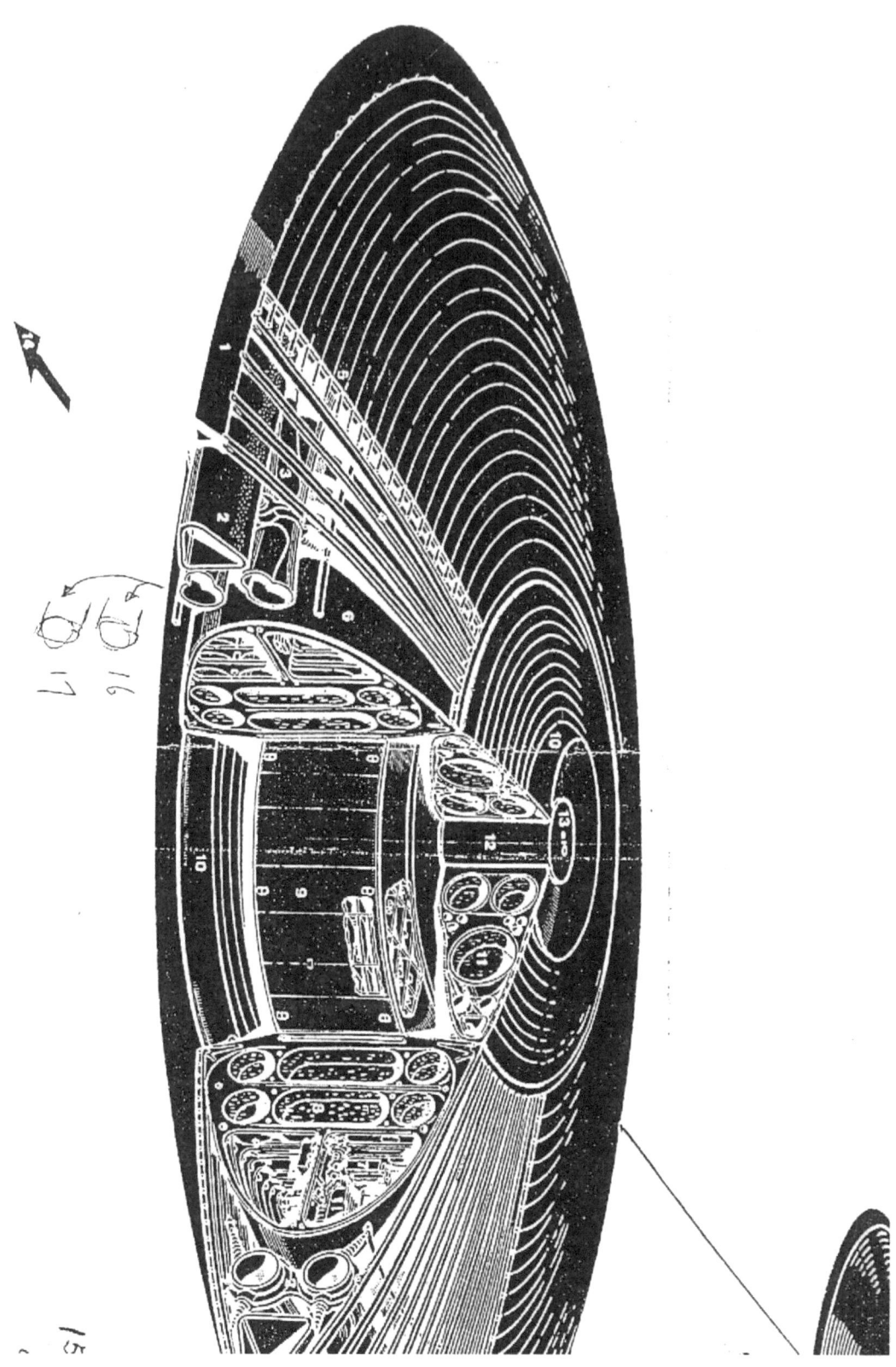

NAZIS ARE MAKING NEWS AGAIN

Late in 2010 headlines in several European newspapers, including Britain's Mail and Telegraph screamed that Hitler's last attempt to win the war could have been a surprise assault on London and New York carried out with giant antigravity flying disks an attack thwarted only by the end of WWII.

Long circulated in the UFO underground, the story received a new lease on life when German science magazine PM reported evidence for the existence of such an advanced and secret program in the German military during the last days of the war.

Now Iron Sky, a new movie (set for release later this year) suggests that the Nazis not only had the technology to carry out such attacks but that, before the war's end, they successfully transferred operations to the far side of the moon where development has continued to this day. In the film, the Nazis attack again in 2018, but this time with enormous bell— shaped flying disks.

Could there be any truth behind such fantastic allegations? Reporter C.C. von Werklaag investigates for Atlantis Rising,

Parts of this article are transcribed by this author due to misinformation and disinformation... Antigravity

'Die Glocke' was built by the Germans as a special project and tested at a secret base in Ludwigsdorf, near the Czech border. It was powered by a radioactive compound labeled Xerum 525 a "red mercury" which author Farrell claims the Nazis might have used to create an atomic bomb (The SS Brotherhood of the Bell), and which scientists Drs. David Clarke and Steve Young associate with "vril power." Dr. Clarke notes a connection between the inner circle of Nazi political power and the Thule Society, a group obsessed with vril, while Dr. Young points out how seriously vril was examined by Nazi for its potential in the manufacture of weapons.

Die Glocke's complex system of opposing turbines was purported to generate a field of opposing turbines was purported to generate a field of antigravity so powerful it wreaked havoc on all life in its vicinity and may even have teleported matter over vast distances. Some speculate that Die Glocke functioned as a time machine, a theory often associated with UFO conjecture and corroborated by former U.S. Army remote viewer Joseph McMoneag1e who claimed in 'Atlantis Rising' #18 (1999) that "UFOs are real vehicles — possibly time machines. The deleterious effects described above sound like the pivotal weapon the Nazis were searching for except their interest in antigravity had a more logical origin. With their Luftwaffe's runways destroyed by Allied bombing raids, the need for aircraft that could take off and land vertically (VTOL) became vitally important. The most sensible assumption indicates 'Die Glocke's capabilities for propulsion rather than its use as a weapon. The Nazis desperation to create such technology may have prompted their scientists to sidestep conventional math. Max B. Miller's article
"Field Theory and Gravity Drive"

(Fate Magazine, Vol. 11, No. 5, May 1958) notes   Dr. Hermann Oberth's conclusion that the "behavior of the UFO discounts any means of propulsion including the reaction rocket — known to us, " and that the "principle of an 'a •gravity device' might be expected." An interesting conclusion since, as Farrell points out, Oberth was directly involved with Die Glocke and the creation of Nazi UFOs.

Analyzing Oberth's pronouncement from the perspective of contemporary physics, mathematician Ward Locke combined Einstein's gravitational theorems with a later tensor model developed by Hermann Weyl. The math revealed the potential for the generation of an antigravity field when the equation begets a negative number. Professor Locke confides that sustaining such a system requires a continued energy input of at least 900 kiloampæres (or the transfer of 1020 electrons) per second. With— out an effective heat—sink, the temperature will instantly rise to 28,000 e Kelvin (nearly equivalent to the surface temperature of the sun). Although dubious that the Nazis harnessed such a force, Professor Locke concedes that, if true, this may have been the reason behind the short intervals sustained during Die Glocke's early tests. The scientists simply couldn't generate enough power to keep it going. It Takes a Rocket Scientist to Understand It

Discussing the wave of UFO sightings occurring in the decade after WWII, Frank Edwards detailed another interesting connection between this phenomenon and the seemingly omnipresent Hermann Oberth.

West Germany... the scene of so many visits by the UFOs, including some reported landings, named the world— famous rocket and space— travel scientist, Professor Hermann Oberth, to head their probe The out— spoken Oberth said...in 1954:
There is no doubt in my mind that... (UFOs) are interplanetary craft of some sort It is also our conclusion that they are propelled by distorting or converting the gravitational field" (Strange World).

Was similar technology channel led into experiments conducted by the U.S. government? Farrell notes the probability that Werner von Braun, along with other German rocket scientists brought over during Project Paperclip, most likely contributed to research which fueled UFO sightings from the late 1940s to now.

Several failures required cover— ups like those at Roswell, NM and Kecksburg, PA.

The involvement of von Braun and Oberth with UFO phenomenon pales in comparison to the identities of those who oversaw Die Glocke. Reichs— fåhrer Heinrich Himmler, head of the SS, was obsessed with the occult, yet his involvement with Die Glocke seems tangential until one con— siders his closest associate. In 1942, Himmler chose Dr. Hans Kammler, formerly a high— level Air Ministry officer in charge of engineering, to take over the rocketry program. In his memoir, Inside the Third Reich, Albert Speer claims that at first he "liked (Kammler's) objective coolness" but later understood how Kammler's "zeal" made him dangerous. Speer states that "SS Gruppenfåhrer Kammler, already responsible for the rocket weapons, was to be in charge of the development and production of all modern aircraft" (emphasis added) forcing him out of this position. Speer notes the incredible number of workers (both skilled and slave labor) commandeered by Kammler for his endeavors. The creation of "All modern aircraft" would have provided Kammler free reign to pursue the elusive 'wunderwaffe' sought by Hitler. The displacement of "half a million workers a year, '1 noted by   273   Speer, would have provided Kammler the means to achieve this goal.

As WWII round to a grim close for Germany, Kammler disappeared.

Kammler reappeared in Argentina, aided by Argentine government. He was subsequently seen in Paraguay and passed away in Argentina at age 88.

From Antarctica to the Moon (and Back Again)

Iron Sky (2011), a soon—to—be— released Finnish film, presents an alternative view of history. While the Allies dug through the rubble after WWII, the Nazis built a secret facility in Antarctica where they developed UFO technology. They soon moved to a base on the far side or dark side of the moon and continued, throughout the decades, to construct an armada of flying saucers. The Nazis plan to invade the world anew in 2018.

The problem with that stupid film is that there is no truth to any of the claims made in this film. The movie's premise seems like much Hollywood—inspired science fiction.

Furthermore, 2018 will bring many natural disasters and earthquakes, seaquakes worldwide as will be witnessed on a global scale These major disasters will be evident on a continual basis from now through the years 2012, 2013, 2014, 2015 & 2018!

The German magazine Faktor—X (issue 11, 1997) reported Apollo 14 astronauts' description of strange objects visible on the surface of the moon. Astronaut Gordon Cooper claimed to have seen UFOs and Neil Armstrong said that NASA was not the first to reach the moon. Ham radio operators, by—passing frequencies used by TV and radio during the 1969 moon landing, heard an exchange between NASA and Armstrong describing other spacecraft lining the crater rim in which the Eagle module had landed.

Furthermore, I maintain that NASA never went to the moon in the first place, as they could not have penetrated the Van Allen radiation belt with their puny rockets and missiles, as was presented to the public from 1969-1974! Readers need to be informed of real truths and not misinformation and disinformation spread by NASA and government agencies! This author has stated the real facts pertaining to German flying saucer bases in Antarctica in his book 'Gravity, Matter & Space Travel', published in 2006!!! This nonsense by others has to stop!

REFERENCES

Note: Some of the page numbers are no longer relevant in this edition of Dr. Petermann's book "God's Works in Our Universe" and in "God's Works in
Mysterious Ways" as the editor was did not incorporate them in this print.

"Law 'G — Richard West Mann page 2.

"Gravity, Matter & Space Travel "   Hans J. Petermann, Ph.D. Trafford

Publishing, Bloomington, Indiana December, 2006

"The Esoteric Sciences", Hans J. Petermann, Ph.D., Volumes 1 + 2 @ 600 pages each, 1999 - 2003— various pages 44 53, etc.

Einstein's theory of relativity is wrong! Prana acts as a

force of energy; Matter is 'similar' to mass; Karmic momentum may ' operate at the speed of light only in our own solar system, as it is variable in comparison with superluminal light speeds in interstellar and "GOD's particles" at intergalactic + interdimensional inter— galactic forces!

More detailed data is available @ 'Gravity, Matter & Space Travel', by this author as mentioned above. Is God's Mind at work in the multi verse, constantly creating inertial momentum and perceptible life forms interdimensionally? Apparently so.

Astrologer John E. Mirehiel — we are all "god lings" in essence. See the 'Harmonic Concordance of 2003. • •

Author Arthur Avalon   ' The Serpent Power

Anodea Judith "Wheels of Life work on chakras and cosmic energies

Dr. V. Fred Rayser, The Golden Rule, 20020
Paramahansa Yogananda the Divine Romance 2003.

Arthur Avalon The Serpent Power' - 2003
The "Taoist Daily Discipline," Master Nie of the College of Tao in Los Angeles, California 2000
Noel Huntley, "Escape from the Universe", 1985

Spectrum news New Times,

http://mars—news.de ,

The Sand Whales of Mars.

A recent 'Cosmic Consciousness revelation', Hans
J. Petermann Ph.D., 2011

Rings of Fire around Earth — Prof. Dr. Hans J. Petermann, Scientist, transcribed from 'Atlantis Rising' Magazine.

Nikola Tesla & the Ether confirm that Einstein's theory of relativity is wrong!

The detailed study of Tesla and his many great break— through inventions still having their impact in energy fields today

Our Inhabited Solar System — portions transcribed from
George Adamski 's books 'Flying Saucers Have Landed'
(co—authored with Desmond Leslie, ' Inside the Spaceships' (Flying
Saucers). Farewell, Inside the Spaceships (1980 reprint includes Adamski's account from 'Flying Saucers Have Landed) of the day prevented further action against another member of the aristocracy.

What is it about Gilgamesh that is recognizable as saurian and never stated? Ostensibly his appearance is described in the first part of the poem where the details of his birth are provided.
The tablet begins with the statement that Gilgamesh is two—thirds divine, being the progeny of a goddess and a priest. The next 4 lines seem to be devoted to describing his appearance. For some reason, however, they have been mutilated. Were these lines deliberately defaced by later officials and priests in order to hide the true appearance of Gilga— The patriarchs, god—kings, priests, generals and other members of the aristocracy were also part saurian, and exhibited certain characteristics which set them apart from ordinary

people probably large patches of scaly skin referred to as "the badge of priesthood". They probably also had horns and chin whiskers.

Perhaps it is these reptilian traits that caused such consternation in Genesis when Noah is seen naked by his sons. The reaction is so completely illogical and baffling that it can only be assumed that Noah was hiding something about his appearance from his sons. Perhaps it was the "badge of priesthood", patches of scaly hide such as was on his brother Nir.

Plate 28 Saturn (Voyager photo), plate 29 a NASA drawing shows the known planets of our solar system.

Gods and Spacemen Throughout History, W.R. Drake, Chicago:
Henry Regnery Co., 1975, p. 41 spacemen in the Ancient East,
W.R. Drake, London: Neville Spearman, 1968, p.65
Open cit., Gods and Spacemen Throughout History, p. 227.
The Expanding Case for the UFO, M. K. Jessup,
New York: Citadel, 1957, P. 136   Science Digest, Sept. 1985
San Francisco Chronicle, June 11, 1979

UFOs — A Scientific Debate, "Science in Default" James McDonald, edited by Sagan and Page, New York: Norton,
1974, p. 56 ff.

Scientific Study of Unidentified Flying Objects, University of Colorado, Bantam, 1969, Case 5: pp. 260 - 266

Operation Mind Control, W.H. Bowart, New York: Dell, 1978, p. 24

Admiral Byrd's Private Diary, Page 260

Flugscheiben liber Neuschwabenland — Page 20

Secrets of the Holy Lance Thunderbird Press 1988 P. 261

Hitler's Ashes Seeds of a New Reich — Page 261,

Thunderbird Press,

1989

Gravity, Matter & Space Travel Prof. Dr. Hans J.
Petermann, Trafford

Publishing, Page 262, 2006

(Please Note: Page references in this book have been deleted or modified and will not apply to any text reference in this book as the publisher has removed all page numbers.)

Antarctica Revisited   Pages 263 - 266

Flying Saucer Mothership — Design by Prof. Dr. Hans J.
Petermann,

Pages 267 - 269

Dedication & Memorial page 1   http://www.south-pole.com/p0000151.htm Operation High Jump — A copy of this is available on VHS format only from this author, by just calling him @ (760) 327— 4761. Cost for this video - $ 30. 00, plus $ 5.00 for shipping & handling by sending payment to; Prof. Dr. Hans J. Petermann, P.O. Box 74, Palm Springs, CA 92263-0074.

Prof. Petermann also has copies available for sale in VHS format only pertaining to a demonstration of 'Brown's
Gas' at the Tesla Extraordinary Science Symposium in Colorado Springs, Colorado 1993 conference + workshop. Prof. Petermann was an associate of the late Yull Brown, working with him for 1 year in California. A copy of this video $45.00 including shipping & handling   His other books 'THE ESOTERIC SCIENCES', Volumes 1 + 2 are available for $ 90.00 including shipping & handling They are 600 pages each...

Dr. Peterman has other books available for purchase like 'Esoteric Curiosities of the Plant Kingdom' are available for $ 45.00 including shipping & handling.

Finally, his Computer disc entitled 'Black Projects @ areas 51 & S4 is available for $ 45.00 each including shipping & handling. This fantastic disc is 2 hours long & packed with fascinating data pertaining to U.S. made 'flying saucers'

• • •
Prof. Dr. Hans J. Petermann shares a website with
Carmen Muller

@ vww. mullerpower. com

The private diary of Rear Admiral Richard E. Byrd can be obtained by contacting

'ADVENTURES UNLIMITED' press + publications at: 1-800-
718-4514. 'ATLANTIS RISING' IS ANOTHER
GOOD SOURCE FOR VERY GOOD SCIENTIFIC PUBLICATIONS: Contact e 1-800— 2288381 to order magazine.

My latest transcription of Richard Benson's data will appear in 2011!

POSTSCRIPT

This author is a native of Vienna, Austria.

He was educated in Toronto, Canada & then in Long Beach and Fullerton, California in the Natural Sciences and German 8 also at the University of California in Los
Angeles, California. Holding M.A. degree in German and M. Sc. from 2 California State Universities in California.

Subsequently he was a Professor with the Los Angeles and Palm Springs Unified School Districts and at the College of the Desert in Palm Desert, California.

In the 1990s he was lecturer at the Tesla Science
Symposiums in Colorado Springs, Colorado from 1992 — 1999. He's the author of "Esoteric Curiosities of the Plant
Kingdom", 'The Esoteric Sciences', volumes 1 & 2   @ 600 pages each. 'Gravity, Matter & Space Travel' was published in December, 2006, dealing with new discoveries concerning planets Mars and Venus and our inhabited Moon. . . All planets and moons have an atmosphere. Ganymede, one of the giant moons of Jupiter, has been inhabited for many years their 'flying saucers' have visited planet Earth for a long time. Prof. Hans J. Petermann also developed a magnetometer for N.A. S.A. many years ago.

He's presently developing a 'magnetic generator' and magnetic motor to be manufactured by year 201 g Prof. Hans J. Petermann shares a website with Carmen Muller @ www. mullerpower.com, originally designed and developed by the late Bill Muller of Penticton, B.C. Canada. He has contributed many research articles on alternative energy projects dealing with clean energies. Recently he also finished work on a CD concerning black projects at area S—4 and area 51 in the 'off—limits' desert area in the Nevada desert.

The author is fluent in 7 foreign languages due to his very extensive travels worldwide.

Private analysts say only powerful explosions can blast human bodies with such a force that they end up on an adjacent building.

In her book Behind the Scenes: Ground Zero— a Collection of Personal Accounts, author Gail Swanson writes that New York Fireman Nicholas DeMasi claims that at least 3 of the 4 airline Black Boxes were found and are under FBI control, which the FBI denies. The FBI has a history of suppressing politically incorrect evidence.

New York City auxiliary fire lieutenant Paul Isaac   Jr. said at the 4th 9—11 anniversary "I know 9— 11 was an inside job. The police and firemen know it was an inside job! " He also reiterated the testimony of a 9— 11 survivor who had said that emergency radios were buzzing with information about bombs being detonated inside the towers.

It's amazing how many people are afraid to talk for fear of retaliation! Researcher Vincent Sammartino, who had also attended the 4th memorial on 9—11—05, said: "1 just got back from Ground Zero. People know the truth. Half of the police and firemen were coming up to us and telling us that they know that 9—11 was an inside job. I had tears in my eyes! "

In a C—Span press conference FBI agent Robert Wright said that he has taped recordings proving that the FBI knew at least 2 months in advance that the WTC was going to be hit in the very same fashion that it was hit. FBI director Mueller had threatened that there would be dire consequences if he opened his mouth.

Attorney Stanley Hilton has filed a class action civil lawsuit against George W. Bush and key members of his administration for mass murder. The law suit, which represents some 400 survivors of 9—11 victims, alleges that the Bush administration was a 9—11 co— conspirator. Hilton claims his evidence includes documents and sworn witness statements from top government officials. He says he has received several death threats.

OpEdNews reported that 7 CIA veterans called the official
9—11 Commission
Report a cover— up. They were part of 25 ex—CIA men who sent a letter to Congress offering their services in a new, independent 9—11 investigation.

John Farmer, Dean of Rutgers University's School of Law, former Attorney General of New Jersey and Senior Counsel of the 9—11 Commission, says in his book 'The
Ground Truth' that the official findings of the 9—11 Commission were based on false testimony and documents and are almost entirely untrue.

Emad Salem, an FBI undercover agent, testified in the trial against Ramzi Yousef and Abdul Hakim Murad that the FBI had foreknowledge of the Feb. 26, 1993 attempt to blow up the WTC. Salem said he was told by the FBI that the agency would allow the plan to proceed, but that it would thwart the attempt by substituting a harmless powder for the explosives. However, the substitution was not made. If it wouldn't have been for a minor strategic misplacement of the bomb, both towers could have fallen, potentially killing some 50,000 people.

Numerous 9—11 survivors with eyewitness accounts that disprove the official version had fatal 'accidents' or committed 'V suicide". Others are afraid to talk.

Outspoken former Italian president Francesco Cossiga said in an interview with Italy's most prestigious newspaper Corriero dell a Sera on Nov. 30th 2007, that 9—11 was an inside job, perpetrated by the CIA and Mossad. "I have always said this was common knowledge among global intelligence agencies. The major media networks are also aware of the 9—11 hoax. The people are always the last to know the truth. " For de— tails do a web search for "Francesco Cossiga".

_____

For additional information and documentaries visit:

http://911dvdproject.com/ ; http://911review.com/ ; http://www.pilotsfor911truth.org/ ;
http://hugequestions.com/ ;

http://letsrollforums.com/letsroll911-org-communitywebsite-t26331.html ; http://www.911truth.org/ ;
http://www.911proof.com ;
http://stj911.org/ ; http://www.journalof911studies.com/

THE SECRET COVENANT

The Illuminati Agenda author unknown

An illusion it will be, so large it will escape their perception. Those who will see it will be thought of as insane. We will accomplish one drop at a time which will prevent them from seeing the changes as they occur. We will keep their lifespan short and their minds weak while pretending to do the opposite. We will use toxins, aging accelerators and sedatives in food, water and air. They will be blanketed by poisons everywhere they turn. We will promise to find cures from our many fronts, yet we will feed them more poisons. The poisons will destroy their minds and reproductive systems. From all this, their children will be born sick. When they give birth we will inject more poisons into the blood of their children and convince them it's for their help. We will teach them that the poisons are good, with fun images and musical tones. We will start early on. When their minds are young we will target them with what children love most, sweet things. When their teeth decay we will fill them with toxic metals that will kill their mind and steal their future. When their ability to learn has been affected, we will give them toxic drugs that will make them even sicker and cause other diseases for which we will give them more drugs. We will render them docile and weak. They will grow depressed, obese and feebleminded, and when they come to us for help we will give them more poison.

We will focus their attention toward material goods and entertainment so they will never connect to anything essential. We will guide them and letting them think they are guiding themselves. They will live in self—delusion. Oh yes, so grand the illusion of freedom will be that they will never know they are our slaves. Their minds will belong to us and they will obey. If they refuse, we shall find ways to implement mind— altering technologies. We will include fear as one of our weapons. We will establish their government with 2 opposites within. We will own both sides. They will perform the labor for us and we shall prosper from their toil. We will make them hate and kill each other when it suits us. We will continue to prosper from their wars and their deaths. We will take over their land, resources and wealth, and deceive them into accepting laws that will steal the little freedom they will have left.

We will establish a money system that will imprison them forever, keeping them and their children in perpetual debt. When they shall band together we shall accuse them of crimes and present a different story to the world for we shall own all the media. When they shall rise up against us we will crush them like insects for they will have no weapons. We will recruit some of their own to carry out our plans.

Those they look up to we will enlist to push our poisons. The recruits will be called "initiates" and we will reward them with money and great titles, but they will never become part of us. When a light shine among them, we shall extinguish it by ridicule or death, whichever suits us best. When our goal is accomplished a new era of domination will begin.

This is the secret covenant by which we shall live the rest of our present and future lives, for this reality will transcend many generations and life spans. This covenant must never be known to exist. If they ever find out that together they can vanquish us, they will take action and no person will give us shelter.

In Defense of Ron Jones Wiki Leaks

Wiki Leak's release of classified information has generated a lot of attention in the past few months. The hysterical reaction makes one wonder if this is not an example of killing the messenger for the bad news. Despite what has been claimed, the information that has been so far released, though classified, has caused no known harm to any individual, but it has caused plenty of embarrassment to our government. Losing their grip on the 'empire' is not welcomed by the neoconservatives.

There is now more information confirming that Saudi Arabia is a principle financier of al Qaeda, and that this should set off alarm bells since we guarantee its Shari a—run government. This emphasizes even more the fact that no al Qaeda existed in Iraq before 9—11, and yet we went to war against Iraq based on the lie that it did. It has been charged by experts that Julian Assange has committed a heinous crime, deserving prosecution for treason. But should we not at least ask how the U.S. government can charge an Australian citizen with treason for publishing U.S. secret information that he didn't steal?

The New York Times, as a result of a Supreme Court ruling, was not found guilty in 1971 for the publishing of the Pentagon Papers. Daniel Ellsberg never served a day in prison for his role in obtaining those secret documents. Yet the release of this classified information was considered illegal by many, and those who lied us into the Vietnam War and argued for the prolongation were outraged. But the truths gained from the Pentagon Papers revealed that lies were told about the Gulf of Tonkin attacks which perpetrated a sad and tragic episode in our history. Just as with the Vietnam War, the Iraq War was based on lies. We were never threatened by Weapons of Mass Destruction or al Qaeda in Iraq. Any information that challenges the official propaganda for the war in the Middle East is unwelcome by the administration and supporters of these unnecessary wars.   Questions to consider:

#1: Do the American people deserve to know the truth regarding the ongoing wars in Iraq, Afghanistan, Pakistan and Yemen?

#2: How can an army private access so much secret information? #3: Why is the hostility mostly directed at Assange, the publisher, and not at our government's failure to protect classified information? #4: Are we getting our money's worth from the $80 billion per year we spend on our intelligence agencies?

#5: Which has resulted in the greatest number of deaths: Lying us into wars, or Wiki Leaks' revelations, or the release of the Pentagon papers? #6: If Assange can be convicted of a crime for publishing information he didn't steal, what does that say about the future of the First Amendment and the independence of the Internet?

#7: Could it be that the real reason for the attacks on Wiki Leaks is more about secretly maintaining a seriously flawed foreign policy than it is about national security?

#8: Is there not a huge difference between releasing secret information to help the enemy in the time of a declared war which is treason and the releasing of information #9: to expose our government lies that promote secret wars, death and corruption?

Was it not once considered patriotic to stand up to our government when it is wrong?

Thomas Jefferson had it right when he said: "Let the eyes of vigilance never be closed. "

The above are excerpts from a recent speech on the House Floor by Congressman Ron Paul. If you are interested in subscribing to his publication FREEDOM REPORT or supporting Dr. Paul 's cause, contact: F.R.E.E., Inc., P.O. Box 1776, Lake Jackson, TX 77566.

Ph: 979—265—3034, ronpaul.org. Listen to his weekly update line at: -888-322-1414.

A PLETADTAN MESSAGE

The Pleiadians are from a star system called Pleiades. This star system is a small cluster of 7 stars located in the constellation of Taurus the Bull. It is some 600 light years from planet Earth. They are a humanoid race (very similar to our human race) who visit Earth often, and with whom we share a common ancestry. Lyrans from Lyra are our common ancestors. The names of the 7 stars in the Pleiades are Taygeta, Maya, Coela, Atlas, Merope, Electra and Alcyone.

Because of the many wars on Lyra, many peaceful Lyrans left on their space-crafts and traveled for many years until they came upon the suitable 7 cluster stars in Pleiades. They landed there and started their own civilization approximately in the year 228,000 B.C. The Pleiadians are a very ancient race of extraterrestrials. They have kept a record of the complete history of Earth's human evolution from the very beginning to our present time. Pleiadians claim our Earth is 626 billion years old.

Around 225,000 B.C. the Pleiadians discovered a small sun system with a planet called Earth on one of their scouting missions away from the Pleiades. They discovered on Earth 3 groups of uncivilized people living there. The larger of these groups were light— skinned and were of Lyran descent. The Lyrans had landed on Earth earlier and were forced to stay on Earth and enter into an incarnational cycle, because of their ill treatment of the original brown—skinned natives who had originated from yet another type of humanoids who had landed on Earth before the Lyrans did. This became their karma. At this time the Pleiadians decided to stay and create civilized societies on Earth.

The Galactic Federation allowed the Pleiadians to enter into an incarnational cycle with humans on Earth. The locations designated for them to do this were Bali,
Hawaii, Samoa and India. Civilizations came and went on Earth with many wars, peaceful cycles and natural disasters. The Pleiadians stayed with humans on Earth until 10 A.D. trying to help develop various civilizations such as Lemur i a, Maya, Inca and a civilization at Machu Picchu. They also tried to guide humans toward a more spiritual path.

At about 10 A.D. the last Pleiadian leader called Plejas left Earth for good, because the Pleiadians felt it was time for humans to evolve on their own. Before leaving Earth, the Pleiadians left a spiritual leader Immanuel, the main character of the composite figure who later became known as Jesus. Immanuel was a very evolved soul, whose parents were Gabriel of the Pleiades system and Mary who was of Lyran descent. Earth continued to evolve on its own without direct Pleiadian leader— ship until about the turn of the Millennium. In the not too distant future, as Earth enters deeper into the Photon Band, the Pleiadians will help bring humans on Earth into the Light.

The following is a brief description of Pleiadian Culture on the home planet of Erra. Erra is a smaller planet, 10% smaller than Earth, and is located near Taygeta. The Pleiadians do not worship a God as we do. They see most earthly religions or spiritual movements as a promotion of dependency, enslavement of Consciousness, and as being in violation with the laws of Creation. They are on a fifth dimensional frequency, which is one of Love and Creativity. About 400, 000 people live on Erra, which the Pleiadians feel is the ideal number for the welfare of their planet. The Pleiadians are telepathic and therefore have no need for telephones. Their means of travel around the planet is done by a tube system. Pleiadians are primarily vegetarians, but on occasion eat some meat. They have no medical concerns because they control their health by using their own mental powers. The average age of a Pleiadian is 700 years. No currency is needed. All material goods are given to the people freely, based on their contributions to their society. When children are born they are taught for the first ten years the purpose of their lives. The next sixty to seventy years they are trained in various skills and occupations. Pleiadians travel in beam— ships which can fly billions of miles in a fraction of a second by traveling in hyperspace. Because of this technology, the trip from Erra in Pleiades to Earth takes only seven hours. Although the Pleiadians are perfectly willing to and able to wage war against hostile intruders from other star systems, they have achieved a period of over 50, 000 years of perfect peace and harmony within their own civilization. Below is a message that has been received from the Pleiadians before the turn of the Millennium:

"We are glad we have the opportunity for this message. This information is important and will be received by many people who have never encountered this type of information before.

Let's start with the idea of government. One of the biggest problems of humankind is that you have abdicated your responsibility to government, be it worldly or religious. Yes, you have been coerced and rewarded towards this abdication, which came about through thousands of years of listening to the wrong people who told you that you are flawed and helpless and lack self— control. So most people have little self— esteem. The highest truth that you can grasp as a human being is that the government outside of yourself is a reflection of your inner government. Your inner government should be your willingness to say, 'The buck stops here. I am responsible. I am in charge. ' But you have been trained for 2,000 years that you cannot trust governing your own self. It is only the 'renegades', the 'black sheep' who decide they are in charge, who heal themselves, who decide not to accept someone's proclamation that you cannot do this or that. So your exterior government is ballooning out of proportion occupying the vacant space of your interior government. When you expect an exterior government to solve all your problems and take care of you, you are in conflict. Part of your spiritual awakening is mastering the ability

to live in peace and harmony, and still have the courage to say, the emperor is wearing no clothes, no matter how difficult the truth may be for others to hear.

You are creating your own reality with your mind, your thoughts. Fears, worries, bad intentions fill the ethers around you and become alive.
Remember, your power ends where fear begins. Your negative thought forms attract lower vibrational entities from the astral Plane, entities that have been around for eons and they possess you. These possessions will eventually be understood, sometime in the future. When the decades around the turn of the Millennium are seen as the darkest of the dark ages, it will be understood that a large majority of

humans on Earth were possessed by lower entities that fed off their vibrations and kept them in fear, doubt, worry, violence, misuse of sexual energy, etc. For some readers these are few big concepts to grasp. Yet we cannot insulate you any longer. You cannot stay in kindergarten forever.

Fear and worry is also one of the greatest detriments to your health.

Loving yourself, the simple act of looking in the mirror and saying, 'I love you' to yourself and meaning it, and being able to receive that love from yourself through the eyes and connecting with it, is a simple act that greatly enhances your immune system. Your ability to say it is my intent as a human being to obtain wisdom, to live with dignity, to live with courage and to manifest what I need. I would like for this knowledge to come to me. I ask to be protected and guided. I ask to be connected to the great Isness, the First Cause. I ask to be in flow with Universal Energy. These are statements of intention of abundance. Abundance is not money in the bank. Abundance is your ability to know that you can create the next moment full of what you want. If your hands are empty and your will, your spirit, your intent, your heart, your energy is full, you will always create what you need. The more the economic system collapses i the greater the abilities within you become manifest!

For a long time, we have said that the greatest mind control machine on your planet is your sacred television. We have suggested if you want to evolve in consciousness, take your television, put it on the curb with a note saying, this is a very effective mind control machine, free to the first sucker. ' You are like dry sponges, you soak up misleading images put before you, and your subconscious mind does not know the difference between what's real and what isn't. We would say that in America some 94% of human beings are walking robots, mind— controlled human slaves who are totally disconnected from their potential.

The more you claim your power, the more things will manifest through you. Eventually, you will be able to pull things into form through dimensions and truly make food come out of thin air. It is not so difficult to do. The stories of the ancient teachers that were compiled into the character of Jesus, who walked on water and performed many 'miracles' were all dimensional shifters. You were created in the image of the great Tsness. You can do the same. But you must have faith. You must have your will, your power finely aligned, like the ruby laser beam. Then you must focus on your wants, and then let go.

However, if you become obsessed with what you want, it will not work.

The reason for living is to serve others, become sovereign, and to take charge, just like Prime Creator. Prime Creator, First Cause, the Isness is waiting for you, not to be worshipped by you, but to be Its equal; and it will graciously acknowledge your greatness.

We would ask your readers to be less gullible and have greater faith in themselves."

Billy Meier, a Swiss intermediary, has been in direct physical contact with the Pleiadians for over 50 years. Nobody was ever able to prove any of his thousands of close—up photos and films to be fakes. He was given physical evidence by the aliens for scientific analysis, with startling results. Dozens of times he has received specific information and predictions that have since proved true. Meier has also survived 19 documented assassination attempts.

Suggested reading: Bringers of the Dawn, and   Earth; Pleiadian Keys To

The Living Library, by Barbara Marciniak. The Pleiadian Agenda, by Barbara Hand Clow.

And Still They Fly, by Guido Moosbrugger. In book stores or from TGS Services at: 903— 876—3256, www.hiddenmysteries.com

Suggested viewing: The Meier Contacts: The Key to Our Future Survival (DVD).

Available from:
www.theyfly.com For additional information do a Web Search for "Pleiadians.

ECONOMIC WOES, A CONSEQUENCE OF 'FREE' TRADE

By Deanna Spingola

The abolition of nationalism and borders under the guise of 'free' trade has been the ultimate Illuminati objective since the late 1700s. The product of the labor of its citizens determine a nation's prosperity. A brisk manufacturing base is essential, augmented by the service industry. Nationalists believe in free but fair trade with reasonable tariffs that protect the nation's wealth. Neocon Republicans along with Liberal Democrats have participated in the legislation of all of the nation's unfair trade agreements.
Officials began negotiating the NAFTA under Reagan in
1986. NAFTA was signed on Dec. 17, 1992 by president
George H. W. Bush, Canadian Prime Minister Brian Mulroney and Mexican President Carlos Salinas, by 1999, due to NAFTA, U.S. job losses amounted to about 600,000, mostly women, Blacks and Hispanics. Wages in Mexico sank by 29%. NAFTA, promoted as a strategy to reduce US trade deficits, actually increased those deficits. In 1993 we had a $ 1.6 billion surpluses with Mexico, and in 2007 we had a $74.8 billion deficits. Mexico, as a US trading partner, is too poor to be an export market for American goods. Over 80% of the American population opposed NAFTA. After NAFTA, the Clinton administration hammered through some 200 additional trade agreements.

Unfortunately, Mexico, along with many other third world countries, became victims of the economic hit men. John Perkins recently revealed their tactics in his 'Confessions of an Economic Hit Man'. Mexico was generating about $ 30 billion a year towards paying interest on its international banking cabal loans. Yet the actual cost amounted to $40— 45 billion per year. Alan Greenspan increased the interest rates in late 1994 which deliberately devastated the peso by about 40%. Congress immediately approved a $40 billion loan despite debt— burdened Mexico's inability to repay it. That loan was merely a transfer from the pockets of taxpayers into the coffers of the International Bankers. The recent bailouts and stimulus packages, disguised as assistance to the populace, again was a huge transfer of wealth from the US tax— payers' pockets into the bankers' pockets.

In March 2010 Rep. Gene Taylor, a Mississippi Democrat, introduced legislation that would require President Obama to repeal NAFTA, the trade agreement that began the de— industrialization of America. In his campaign speeches Obama opposed NAFTA but now is negotiating with South Korea, Panama and Colombia to implement additional 'free' trade agreements with those countries. Recently U.S. officials also began trade talks with

Australia, Chile, Vietnam, etc. in what would be the Asia— Pacific regional 'free' trade agreement. Ian Fletcher wrote, "Free trade is bleeding our economy and preventing it from returning to true health. Nobody in the Obama administration wants to talk about it, because as soon as one seriously scrutinizes this doctrine, one begins to discover that 'free' trade may be the biggest myth in American economics. "Donald Trump said, it's unbelievable how easy it is for other countries to get into our markets, and how difficult it is for us to access foreign markets. I have friends in China and OPEC. They tell me in private, 'It's just amazing what we get away with. ' Your leaders are very stupid'.

CONTINENTAL CRACKS

To those of us who love them, maps are wondrous things; and just looking at a map with eyes and mind wide open can allow us to see patterns previously unnoticed.

Looking at a map of North America reveals the ancient, eroded Appalachians in the east and the younger and higher mountains in the western third of the continent. In between is the deceptively flat and seemingly stable heartland its smooth contours broken only by scattered ranges even more ancient than the Appalachians, now so worn down that they are really little more than hills. Dominating the heartland is the vast Mississippi River system with its 3 main branches, the Ohio, the Mississippi itself, and the Missouri, all draining the continent into the Gulf of Mexico in the south from the higher ground in the north. But, in the far northeast is a most peculiar feature, the broad St. Lawrence Valley. This river's flow is far less than that of the Mississippi, but rather like the Rio de la Plata in South America, its lower reaches are abnormally wide far wider than the Mississippi. Since rivers and lakes, for obvious reasons, follow the low ground, it is possible to trace the areas of low elevation along the St. Lawrence, then along Lake Ontario and Lake Erie, the Wabash River Valley, the lower Ohio River, and on down the Mississippi. But the lower regions of the St. Lawrence look as if some great force was trying to split the continent apart.

From all of this, a somewhat ominous pattern emerges. Most well in— formed observers know of the series of powerful earthquakes near the Mississippi and New Madrid, Missouri, in 1811 and 1812, and the countless smaller quakes that have continued to shake that region to this day. Geologists have discovered in this area something they call the New Madrid Seismic Zone and, just to the southwest, the Reel Foot Rift. In fact, there are a number of ancient rift zones, or long faults, in the heartland; and there have been earthquakes almost everywhere in America, including Arkansas, and especially recently, a good many in Oklahoma. There have been earthquakes in Indiana and even on the flat Gulf Coast Plain areas of Texas,

Louisiana, and Alabama where sediment is miles deep. In fact, geologists speak of the St. Lawrence Seismic Zone and the Wabash Seismic Zone; and some consider the New Madrid area to be an ancient rift where, long ago, seismic forces tried to split the continent. Some believe that the rifting began some 750 million years BP (before the present) during the Neoproterozoic Era when North America was part of the continent of Rodinia. Others believe that the rifting may date to the time when the Appalachians formed. Most believe that it is a failed rift. but is it?

Retired Texaco geologist Jack M. Reed finds the conventional view very suspicious. In fact, Reed believes that the entire Gulf of Mexico is an active seismic zone and, possibly, a separate, south— moving crustal plate. He thinks that there may very well be one continuous rift from the mouth of the St. Lawrence to the Mouth of the Mississippi and that the seismic activity in the Gulf has helped to create this rift.
He points to such tectonically active formations as the Monroe Uplift, the La Salle Arch, and the Sabine Lake area on land, and, in the Gulf of Mexico, the Desoto Canyon, and a domed uplift nearby, the Cretaceous Shelf Edge, the Suwannee Strait, and the West Florida Escarpment. The Suwannee Strait is a depression extending across north— western Florida and parts of Georgia, which Reed suspects is connected to the active seismic zone just inland from Charleston, South Carolina a city that experienced a powerful earthquake in the nineteenth century.

Geologists and geophysicists currently tend to believe in the theory of plate tectonics, and if we examine this theory and then, in light of it, take another look at the overall structure of North America, we can at least begin to understand what is happening. Although the exact mechanism driving plate movement is poorly understood, and despite the fact that almost anything that anyone says about it will be an oversimplification, the theory is supported by massive evidence and explains a good deal about the earth. Basically, geologists believe that the crust and the solid upper mantle of the planet are broken up into a number of plates and their boundaries are either divergent: along spreading zones (most of them under the ocean) where magma oozes up from the lower mantle; or convergent: with one plate subducting under another; or transform: where the plates rub against one another horizontally. An example of a transform boundary is the San Andreas Fault which would be an ordinary strike— slip fault were it not considered to be the boundary between the North American Plate and the Pacific Plate. The plates of the lithosphere move over the partly melted layer of the mantle known as the asthenosphere.

Not all geologists agree on the number of plates or their exact boundaries; sometimes it is difficult to tell if a break in the crust is truly a plate boundary or simply a massive strike—slip fault. Also, the boundary is often not a simple line but a wide zone. It was at first assumed that sea floor spreading explained the movements, with magma welling up along the entire length of a rift zone due to convection currents. However, although there are hot spots where convection currents drive magma plumes up through the crust (examples are Iceland, the Big Island of Hawaii, and Yellowstone, among others) it is a bit difficult to explain how a convection current would form a long, narrow line. In the case of Iceland, the hot spot is also on a spreading zone. Perhaps the plates are not only pushed at spreading zones, but also pulled by cooler, denser basalt sea floor sinking under the lighter granitic rocks of continents. Tidal forces and the earth's rotation may also play a role, and perhaps even asteroid and comet impacts have had something to do with it. But the movements, for billions of years, have split continents, moved them apart, and brought them back together, combining and recombining our planet 's surface features. The original nucleus of North America was the North American or Laurentian craton, whose highest portions today make up the Canadian, or Laurentian, shield. It was formed by eruptions of sialic (containing relatively high amounts of silica and aluminum) igneous rock called plutons which were less dense than the underlying mafic (less silica and more iron) rock of the mantle. This formation is related to similar formations in Greenland. Through an imperfectly understood process, much of it was apparently uplifted to heights dwarfing most modern mountains, for, today, after several billion years, there are still very high mountains in Greenland and in north— eastern

Canada 's Arctic islands. This is why the Mississippi flows from north to south and why there still ranges of very high hills in parts of central Canada, Minnesota, and Michigan.

   Most of the high— lands of eastern Canada, including those around the St. Lawrence Valley, are part of this original craton.

Over a billion years ago, a volcanic plateau formed, and was later eroded and mostly covered by sediments, forming today's Ozark Mountains (really just hills) of southern Missouri and northern Arkansas. Another early plate collision formed the Wichita Mountains of southern Oklahoma, perhaps 500 million years ago; and, about 300 million years BP the South American plate is believed to have subducted under North America forming the Ouachita Mountains of Arkansas and eastern Oklahoma and the Hill Country of central Texas; the eroded mountains in between are mostly covered by sediment. This Ouachita orogeny also produced the northwest to southeast trending Southern Oklahoma
Rift.

Some 480 million years BP North America collided with what are now Europe and Africa, crumpling the crust and forcing it up to form the Appalachians, probably about as high then as the Himalayas are today, and forming the continent of Pangea. Later, around 130 million years BP and along roughly the same boundary, the continents separated along what is now the Mid—Atlantic Ridge, forming the Atlantic Ocean. Geophysicist J. Tuzo Wilson noted the tendency for the crust to break, reform, and break again in the same areas; this is referred to as the Wilson Tectonic Cycle. Some contemporary geophysicists, like Tony Lowry and Marta Perez— Gussinye, suspect that this happens because these regions have more quartz in the rocks that weakens the crust. For a long time, the immense North American Plate moved roughly west colliding with the Pacific Plate, which mostly subducted under our continent. New mountains were fold up, volcanoes erupted, and 'I terranes" or islands and fragments of continents accreted to the western edge of North America. All of this gave us the Rockies, the Sierra Nevada and Cascades, the Coast Ranges, and all the other high western peaks.

Now the North American plate, or at least its western edge, appears to be moving to the southwest, while the eastern Pacific Plate moves to the northeast, making the plate boundary, the San Andreas Fault, a strike slip transform fault. But things are not quite that simple. Roughly parallel to the dreaded San Andreas are innumerable other fault like the Hayward Fault along the Great Valley Fault along the eastern margin of the Coast Ranges. The Sierra Nevada (rather like the Tetons of Wyoming or the Sawtooth

Mountains of Idaho) is a huge block of the crust, broken and thrust upward by a fault on its east side and tiltin up more gently on the west. Another look at the map reveals the Gulf of California, or Sea of Cortez, where an extension of the East Pacifi Rise (a seafloor spreading zone) is literally splitting Baja California away from the rest of the continent. The splitting does not stop at the water's edge, but extends, under sediments, under California's Imperial Valley. There are volcanoes all along the Sea of Cortez and cinder cones and geologically recent lava flows throughout south— eastern California. Then there is Death Valley, resembling the Dead Sea in the Middle East or Lake Baikal in Siberia, only without the water. Death Valley, at its deepest point, is 282 feet below sea level with deep sediment under that. Like any sea floor spreading zone, it is bounded by mountains: The Amargosa Range to the east and the Pana— mint Range to the west. North and west of Death Valley, along the eastern Sierra fault zone, are the Long Valley Caldera, site of an ancient mega— eruption, and Mono Lake,
Mono Craters, and the Mammoth Mountain volcano.
North of Death Valley the faulting and low ground extend along the Walker Lane that goes as far as Oregon. So it is not clear that the San Andreas is really the true plate boundary; and eventually, the Sea of Cortez may extend further and further north. And there is at least one other major rift zone that looks like more than just a regular earthquake fault. The Rio Grande rift zone extends from Colorado to El Paso and is bounded by steep faulted and folded mountains to the east, like the Sangre de Christo, the Sandi as, the San Andreas Mountains, the Organ Mountains, and the Franklin Mountains On the west are volcanoes and geologically recent lava flows, like the San Juan Mountains and the Jemez Mountains. There is no way to know if this rift will grow larger or not.

Returning to the heartland and its apparent rift zone, the big problem in predicting what may happen is the fact that we do not fully understand the forces acting on the North American plate, which, rather than simply moving to the west, may now be rotating in a counterclockwise direction... or, at least, its western portion is.

Clearly, if a plate is being pushed in one direction and/or pulled in another, every portion of it, dragging over the asthenosphere, will be affected. If the forces are acting to pull the continent apart along the ancient rift zone in its center, rest assured that it will, indeed, come apart. Remember that high mountains rise today where once there was ocean floor, and level plains and ranges of hills are all that remain of past mountains. The only constant, as they say, is change.

And if change comes to the heartland, will it be imperceptibly slow or catastrophically rapid? Geology, early on, was dominated by Bible— inspired catastrophists; later, uniformitarian ism became the ruling paradigm. Our understanding of asteroid and comet impacts, mega—volcanoes, and mega— tsunamis has moved most geologists more toward the middle, and they tend to believe that generally slow processes are occasionally interrupted by truly awesome events. Yet it would seem that, even if there are massive earthquakes along, say, the New Madrid seismic zone, or even if the land subsides, or the St. Lawrence River becomes even wider, it is unlikely that the entire continent could split open at once or even in a few centuries. And more active areas like Africa's Afar Triangle or the Sea of Cortez here in North America would seem more likely candidates for sudden change.

But sudden earthquakes have become much more frequent in recent decades, and, as the earth shakes its way toward 12/21/2012, we should probably not be surprised at anything!

The last part of this book is dedicated to my dear late friend Phil Schneider, who survived an underground confrontation with 'aliens' at Dulce, New Mexico in 1979. Phil Schneider was murdered by military government agents at his home in Oregon in 1995. MAY HE REST TN PEACE!

In late 1979, there was a confrontation over weapons. A lot of scientists and military personnel were killed. The base was closed for a while • But, it IS currently active.
Note:

Human and animal abductions (for their blood and other parts) slowed in the mid— 1980s, when the Livermore Berkeley Labs began production of artificial blood for Dulce.

The late William Cooper stated: "A clash occurred wherein 66 of our people, from the National Recon Group, the DELTA group, which is responsible for security of all Alien— connected projects, were killed. The DELTA Group (within the Intelligence Support
Activity) have been seen with badges which have a black Triangle on a red background. DELTA is the fourth letter of the Greek alphabet. It has the form of a triangle, and figures prominently in certain Masonic signs.

Each base has its own symbol. The Dulce base symbol is a triangle with the Greek letter "Tau" (T) within it and then the symbol is inverted, so the triangle points down.

The Insignia of "a triangle and three lateral lines" has been seen on

'Saucer (transport) Craft," The Tri —Lateral Symbol. Other symbols mark landing sites and Alien craft.

THOUSANDS OF MILES OF TUNNELS EXIST BENEATH THE
U.S.

WHERE ALIENS EXIST

"This Awareness indicates that there is an awareness by the U.S. government of these caves and caverns under the surface of this Earth. Many are in the United States. Please refer to my first volume entitled "GRAVITY, MATTER & SPACE TRAVEL" for extensive details explaining this scenario. This Awareness indicates that there is in fact an organization called the '165 Mile Club' which is made up of entities working with the government who have been 165 miles deep into the subterranean caverns. This Awareness indicates that TT wishes to warn entities to stay out of these caves and caverns, as there is   grave danger in these, particularly in those which are remote and not visited by tourists. This Awareness indicates that generally those which are known as tourist attractions are not used by these sub-terranean beings.

'This Awareness indicates that essentially the Earth's surface is catacombed with many tunnels, that there are also areas where there is much hot lava these areas are not used by these beings and the tunnels avoid contact with such areas. This Awareness indicates that there are tunnels running about 4,000 miles from the Eastern part of the U.S., to an area in Arizona and into areas north to Vancouver, approximately 1,800 miles. This Awareness indicates there are many other branches and caverns coming off these. This Awareness indicates that there are also tunnels in South America, Tibet, in various places in Europe these running underground. This Awareness indicates there are even tunnels which move beneath the ocean floors, linking the continents together.

DETRIMENTAL ROBOTS (DEROS) LIVE BENEATH THE EARTH

"This Awareness indicates that these so—called Deros are also the basis for certain folklore such as gnomes and elves, goblins, trolls. This Awareness indicates that the folklore relating unto these entities has changed somewhat and that many of the so— called gnomes, trolls, elves and imps that many of these are now projected as being nice little fellows. This Awareness suggests that this is a way of reconciling, or healing, or diffusing the fear and hostility and so—called evilness of these entities, so that they may cease to be the garbage can for human hostilities, fears and enemies.

'This Awareness indicates that those so—called Deros, or Detrimental Robots, those entities live beneath the crust of the Earth — this Awareness indicates that there are many types of entities beneath the Earth; that those closer to the center itself are of a highly evolved nature, whereas those who are sub—surface entities, these are somewhat demented by the effects of living underground for so many thousands of years. This Awareness indicates that those Deros, or whatever name entities desire, are human beings who have regressed through misuse of power, through excessive and continued self— indulgence, and through greed, lust, competition, and joy in sadistic express ions.

"This Awareness indicates that occasionally these entities surface, particularly -through certain buildings which are often built in a way to allow an opening into the subterranean levels, and occasionally these entities walk around on the streets of your cities, and occasionally these entities have even been known to kidnap others. This Awareness indicates these entities are not astral beings, are not entities who are being imagined or experienced after death. These entities are as real as you, 'and in this particular dimension of reality which is yours." This Awareness indicates that these entities are tools of the Alien Force that the Alien Force expresses through these entities, that they are essentially slaves of the Alien
Force, as also are many of the creatures known as
Bigfoot, or of the similar types of hairy monsters."

BIGFOOT ENTITIES WERE ONCE HIGHLY EVOLVED

"This Awareness indicates that these Sasquatch—type entities do not care to move too close to humans on the outer surface of the Earth. This Awareness suggests these entities having great karma to work off, yet fully aware of their karma and very lonely entities. This Awareness indicating these entities having a type of consciousness which allow them to know precisely where any other human is within a 2-mile area of their being and can move away from these humans very easily through that psychic understanding.

'This Awareness indicates these entities as having extremely high beings in that civilization (on the planet Maldek which was destroyed), having great responsibility and misusing that responsibility, and in that misuse did lead to the devastation of that planet. This Awareness indicates this occurring about 300,000 years ago.

This awareness indicates that the movement of the entities through states of reincarnation: as these souls moved into the Earth's sphere, these entities were outcast by the rest of consciousness upon this plane and those inner planes within the Earth's sphere; and these entities do live alone with some communication relating unto beings who have greater understanding and who can work with these entities.

"This Awareness indicates that 85% of the souls upon this plane have lived on Maldek and this is the Luciferian consciousness which is being worked off as part of the Earth's karma. This Awareness indicates this as being dissolved. This entire Illuminati, money— changing trip is but a replay of the memory of the action which occurred on Maldek wherein beings were categorized, numbered and enslaved. This Awareness indicates that the movement which began 60,000 years ago in Atlantis was but a recollection of certain states which began on Maldek.

"This Awareness indicates that it shall eventually become necessary for entities from this plane to release their hostility and their sub— conscious fears of these entities, and in this action the karma that has been held by these entities shall begin to dissolve, and the lamb shall lie down with the beast and the beast shall find a place of rest in the arms of this Awareness."

MANY UFOs COME FROM BENEATH THE EARTH

"This Awareness indicates that many of the UFOs are from this sub-terranean level and are piloted by these Deros, or by other mechanical creatures such as the synthetics, which are made from parts of entities or animals such as cattle. This Awareness indicates that the cerebral and nervous systems of cattle are often put into these synthetics and the synthetics then are given life, and are used, are inhabited by astral beings who enter these bodies and serve as owners for those bodies and receive their programming from the Alien Force.

'This Awareness indicates those entities known as MIBs, or Men in Black, which often show up after UFO sightings these entities also are from this lower region, that these entities also assist the Alien Force. These entities have the capacity to move from one dimension to another, may enter into a physical body from an astral level. This Awareness indicates that these entities are also connected with that which has been described previously as the Illuminati, and that which is the Anti—Christ, and this shall eventually be made more clear to entities as the information surfaces on many levels.

 This Awareness indicates that there are very sinister forces working upon this plane, and within the caverns beneath the surface of the Earth. These forces do need to be understood, but these forces must be understood in a manner whereby they may be disarmed, rather than feared. That it is possible to change the basic motivation of these aliens, that they may rejoin the ranks of the merciful.

AN ANCIENT TUNNEL SYSTEM

There is an ancient tunnel system beneath the earth that literally circles the globe.

This system has existed beneath our very feet for literally thousands of years and very few of us know about it. And those that do have often found their lives turned into a living hell when they dared to tell others of its very existence.

The tunnels radiate outward from the Arctic and Antarctic in every direction and cover every continent on the planet. They were constructed by a civilization that existed before the "great flood". This civilization came even before that of Atlantis, though the Atlanteans later improved on this existing system, adding to it, as well as establishing underground space ports for visitors arriving from other planets who came here in those "early days" to establish friendly "trade relations" with our planet's peoples.

Sometime in the distant past after the collapse of Atlantis — these caverns were at first abandoned and then taken over by negative space aliens, who collabo— rated with a race of our own underground beings (known as the Dero) to rape the planet of all its vital resources. Humans were taken as "slaves" to work in the underground tunnels from which few were ever able to escape. We are told that cities actually exist beneath the Earth's outer crust, and they can be reached by entering and exiting through concealed openings in various locations. Most "shaftways" to these inner Earth cities can be found in remote areas, while others are in more populated areas. There is supposedly even an opening some— where in New York City in the vicinity of midtown Manhattan that can be reached through an abandoned elevator shaft that only a very few know about for obvious "security" reasons.

CONFIRMATION RECEIVED

Back in the late 1940s and early 50s a channel by the name of Mark Probert was very popular on the West Coast. Much of the information he received came from the Yada Di Shi l ite, a 50,000-year-old Tibetan master who was known to keep a "watchful eye" on happenings on both the physical as well as ether ic planes. The Yada confirmed a great deal about the reality of the underground world and even spoke of a race of "Serpent People" who once overran this planet, having come here from Venus in the remote past. Says the Yada:

"They abandoned the Earth because conditions here were not favorable to them." They were of great size and had scaly bodies and large frog eyes, and were very advanced mentally. Morally they were not evolved, but were extremely cruel and vicious. They are still to be found in the interior of Venus. The Venusians of the present day, however, are not descendants of the early type.

"Venusians" who are visiting your Earth at present want to bring peace. They have no desire to occupy the Earth.
There is no present warfare between them and the
Martians

"Your present danger, mitigated for a time by the Guardians from Venus and elsewhere, lies in the progressive breakdown of the upper ethers, that is, of the ionosphere (which regulates the climate of the g lobe and shields us from the impact of the cosmic rays) "Most of what is said about the tunnel system is correct. These were constructed by Atlanteans, partly for communication, sometimes in connection with the search for metals or ores, but chiefly in order to escape the extreme solar radiation and various bacteria from the surface of the globe. The great plague that visited England (1666 A D.) while partly due to unclean living, was mostly caused by these same bacteria.

'The tunnels themselves were not primarily designed for underground living; but in many cases they lead into vast caverns, natural or artificially hollowed out, where a great number of persons spent all their lives. It is true that the tunnel opening under the pyramid of Gizeh in Egypt leads into the caverns under Tibet. As to size, a common diameter of these tunnels was about 150 feet.

"They were constructed by mechanical means or chemical means; that is, by the application of superheat deteriorators. It was like burning their way through the Earth.

Yes, it resembled the heat of the atomic bomb — if such heat could be controlled and directed. It was a very dangerous process; for if the heat blast went out of control the whole globe might suffer most serious consequences. The process was very rapid. The molecules disintegrated. The substances of the Earth are very porous and the gasses were largely absorbed by it.

"In some quarters much is still known about these tunnels. The two great religious Hierarchies of your plane have such knowledge, and in fact have stored great supplies of food in underground depots. They are aware of the approach of their twilight hour. "So the vast tunnel system dating from Atlantean times, with its hub under a still— lived—in city in the Antarctic, is known to the lamas and priests of Northern Buddhism, and to the priesthood of the Roman Catholic Church; it is not only known but made use of; and apparently their soothsayers have foreseen a great surface catastrophe, such as that which might be caused by a Polar Flip, and have made reparations accordingly.

'The one God of the world of matter is the god of Change, concluded the Yada. "All things move into and out of Manifestation in the 3 dimensions. Where does form come from? Mind built it and will change it according to the need for change. The object lost to three dimensions still holds to form another plane. To know this and know that we know it, is to escape much sorrow when forms of cherished things, beloved persons vanish from the world of our sense perceptions."

The above communication from the Yada was first published in the BSRA Round Robin journal of borderland research for March—April, 1953, edited and published by Meade Layne.

TUNNELS OF SOUTH AMERICA

In Southern and Central America as well as in Mexico, the ancient people did not deny the presence or existence of subterranean caves, chambers or tunnels. An examination of the religious beliefs of all these ancient civilizations will reveal this.

The Aztec of Mexico had their dark, dreary and much feared "Tlalxico" which was ruled by "Mictlan, " their god of death. The Mayas of Yuca— tan held a belief in the existence of nine underworlds. These they termed "Mitlan" and they were icy cold as are most subterranean chambers or tunnels. (For proof visit a large cavern in summer clothes and see how uncomfortable you are.) These underworlds were presided over by "Ah Puch, " the Lord of Death. We also have mention of the underground in the Mayan sacred writings, the "Popol Veh;" as well as in the Book of Chi lam Bal am of Chumayel. Even some of the codices seem to refer to them.

Peru and Chile, when they were ruled by the Incas, also reveal know— ledge of the underground. A legend of the first Inca "Manco Capac" related that he and his followers, the founders of the Inca realm, came from underground caves. While the people of the time revered snakes because of "Urcaguay" the god of underground treasures. This god is depicted as a large snake whose tail has hanging pendant from it the head of a deer and many little golden chains. Even the "Comentarios Reales de los Incas" of Garciliaso de la Vega, hints at the existence of the subterranean.

References to the tunnels have come down to us from information that the conquistadores obtained. From some unknown source they had gathered information that the wealth of the Incas domain was stored in a vast underground tunnel or road. Pizarro held the Inca Atahualpa prisoner in order to obtain his wealth. Wealth which it was rumored was stored in a vast subterranean tunnel that ran for many miles below the surface of the Earth. The Inca, if he had the information regarding the entrance to this tunnel, never revealed it. The priests of the Sun god and the Inca's wife determined, it is asserted, the eventual fate of the Inca, by occult means. The knowledge that Pizarro did not intend to spare the Inca Atahualpas life caused them to seal up the entrance and hide it so well that it has never been found to this day.

A few Qui chua Indians, who are pure descendants of the line of priests still have the knowledge of the location, of the entrance to this tunnel. They are the appointed guardians of this secret, at least so it is rumored today in Peru.

Harold T. Wilkins, author of 'Mysteries of Ancient South America', also researched and inquired about the tunnels until he was able to conclude the following: 2 underground roads leave the vicinity of Lima, Peru. One of these tunnels is a subterranean road to Cusco, almost 400 miles to the east. The other runs underground in a southern direction for over 900 miles to the vicinity of Salar de Atacama. This is the large salt desert in Chile. It is the residue of the ocean water which was landlocked during an upheaval of the Earth. The upheaval or cataclysm which created Lake Titicaca and raised Huanaco high above its place on the shore line, creating the Andes and a new shore line on the west coast of South America. (For information regarding this event, see the section titled "Tiahuanaco in the Andes" in Immanuel Velikov— sky's 'Earth in Upheaval.')

The Cordellerias Domeyko in that section of Chile very evidently landlocked a great portion of the sea when it was raised. After the sea water evaporated the vast salt waste, which is almost impossible to traverse, was left.

The tunnel, which has an entrance somewhere in the Los Tres picos triangle, is also said to have a connection with this long southern underground road.

In conjecture that any continuation of southern tunnel   was broken during the cataclysm which created the Andian mountain range. Such a continuation would have connected these ancient tunnels with the reputed Rainbow City center in Antarctica.

AN ENTRANCE TO THE CAVES TN STAFFORDSHIRE,
ENGLAND

This entrance to the caves now lost, was discovered by a laborer, who was digging in a lonely field somewhere in Staffordshire, England.

The exact location of the field and entrance has never been discovered.

However, the story is related in 'A History of
Staffordshire', by Dr. Plot, who wrote the book in the late 1770s.

I shall relate the highlights of the events as they appear in this ancient historical book.

A dull, ignorant laborer was digging a trench in a field which lay in a valley surrounded on almost all sides by woods somewhere in Stafford— shire. The sun had gone down and the laborer, who later related his experience, asserted that just as he was about to stop work, his pick hit a large flat stone. The stone had an iron ring mounted in it. He stayed and cleared the stone, which was in the form of a large oblong. His attempt to pry the stone up was met with failure, but he utilized ropes he had brought to obtain more leverage and managed to slide the stone over. This revealed a stone staircase, which sloped down into the Earth. Since his first thought was that this might lead to an ancient tomb containing treasure, he gradually descended the stone stairway.

While looking back, he could still see the sky glowing with the last light of the sun. His descent continued until he was about, according to his estimation, 100 feet underground. It was at this point that he suddenly reached a sort of landing.

The planet Venus had risen by this time and was shining directly down the shaft, so that he was able to discern the beginning of another stairway, which descended at a right angle to the first one. The possibility of a treasure in gold or jewels, made him feel his way in the darkness down another 120 steps. At the foot of these steps was a turn, and far below down another long flight of stone steps, he could see a pale but steady light.

While descending this long flight of steps, he heard the sound of some sort of machinery or the rumble of a large vehicle somewhere far in the distance. He paused, frightened, but the sound was gone and in the surrounding stillness, he forced himself to go forward toward the light which glowed unnaturally in the bowels of the Earth. Reaching the end of the steps, he found himself in a large stone chamber, the roof of which seemed far above him and the walls of which he could not see even by the light of a globe, which glowed on the floor before him. Suddenly a hooded cowl ed figure appeared from some side passageway. This being pointed what he described as a baton—like object or as we would understand it, a tube — at the light and destroyed it with a thunderclap, which echoed and reechoed through immeasurable subterranean passageways.

The frightened laborer could not remember how he got out of the tunnel or up the stone stairway, when he related his story. Any attempts to get him to visit the valley again were unsuccessful. Others who searched for the digging were unable to locate either because of the terrible wind and rainstorm which occurred that night. This ripped and washed the vegetation of the valley away, leaving only the bare earth with no trace of the trench, the stone or the staircase. Because of this, the
Staffordshire entrance has never been located again.

THE INTRATERRESTRIALS

Reinhold Schmidt was a quietly—spoken, rather self— effacing man who lived in an apartment block on Franklin Avenue in what is known as the Hollywood Flats district of Los Angeles. In 1958 he made headline news with a story more startling than any publicist in the film industry could probably have dreamed up. For Schmidt told the press, radio and television that he had been taken in a UFO on a journey to the Arctic and met a group of men and women who lived, he said, somewhere beneath the North Pole.

Reinhold Schmidt's 'contact' is therefore of especial interest because of its comparatively recent date and because it provides further evidence of the origins of UFOs. It was in August of 1958 that he first told an LA newspaper about his journey and afterwards repeated the story for the rest of the media and the UFO1Ogists who came to talk to him. Schmidt never refused an interview and convinced everyone he spoke to of his sincerity. One of them, reporter Charles Longcroft of the Los Angeles Examiner, wrote later, 'This was the first time T have ever been face to face with someone who claims to have contacted space men or to have been inside a saucer. My impression is that the man has definitely seen something and is not making the whole story up as a publicity stunt.

The facts about Reinhold Schmidt and his journey to the North Pole are these. He was born in 1920 in Hamburg, from which his parents fled after Hitler's rise to power. The whole family came to California and were granted US citizenship. Reinhold was educated in Los Angeles and followed his father into a small office supply business in the city. He became interested in stories of UFOs, he said, after reading Frank Scully's ground— breaking book 'Behind the Flying Saucers' in 1950, but never expected to see one himself. All that changed for him during a few days in August 1958, as he told UFO Report:

I had what I thought was a dream telling me to go to a quarry in Bakersfield. In fact, I learned later it was a message from the space people. On August 14, I drove my Buick to the quarry and after sitting around for several hours, this circular silver craft came down from the sky. It seemed to be made of something like aluminum and access to it was by sliding doors and a ramp that was lowered to the ground. A figure appeared at the doorway carrying some kind of torch in its hand. This was flashed on me and it paralyzed my physical movements for a minute or two without impairing my ability to think or talk. The figure with the torch was then joined by two others. They came down the ramp and I suddenly found could move again. They escorted me into the craft. Rather than leave my car in the quarry, one of the people indicated I could take it in the machine and it was driven up the ramp and brought on board. We then took off and flew north towards Alaska and after that up to the Arctic and over the polar regions. The crew consisted of four men and two females, Schmidt said. They were all tall, with noble features and dressed in grey, one— piece, skin—tight costumes. The women were especially pretty, in the classical style he always associated with paintings of the ancient Greeks and Romans. They spoke to each other in what he recognized as 'high German' because he had been taught the language by his parents. Throughout the entire journey, however, he was always addressed in very precise English.

What was remarkable about the space craft was that from inside the whole of the hull appeared to be transparent and it was possible to see out from anywhere except those sections which were obscured by machinery or control panels. There were couches, chairs and small tables, and although they did not appear to be fixed to the floor, they never moved even when the craft made the most complex and high speed maneuvers. I was able to see all the journey up to the Arctic Circle. When we got there we seemed to go under the Arctic Ocean and enter a huge hole. T was conscious of passing over a strange landscape though we never actually landed.

Reinhold Schmidt said that his 'hosts' never told him precisely where they came from, although he became increasingly convinced their home— land must be somewhere in the region of the pole. It was evident from the flying machine that they had a highly mechanised society, and from their manner he guessed they enjoyed comfortable and peaceful lives. If their mission had a purpose, he felt it was to observe man— kind and stop us from destroying the planet.

Schmidt's voyage lasted for five days, during which he slept several times. He remembered having the sensation of seeing a land not unlike the earth which was lit by a glowing sun rather different from our Sun. He also twice had the impression of crossing a large curve of ocean when the horizon dipped and fell and then righted itself.   On 18 August he was returned to California and once again found himself in the Bakersfield quarry with the Buick beside him. As he stood watching the spacecraft disappear rapidly northwards, he was suddenly aware that he had an unexpected souvenir of the journey that might help him convince people of the truth of the extraordinary events he had experienced. For he noticed that the paintwork on the upper sur— faces of his car had been turned luminous.

One more report brings our history up to date. According to an article in The National Enquirer of 25 February, 1992, a Danish scientist and explorer, Edmund Bork, had recently returned from leading an international team of explorers through the North Pole opening the previous summer. He had apparently found the 1,400 mile—wide opening thanks to studying the ESSA—7 satellite photograph. The Dane told one of the newspaper's reporters:

There's a hole in the pole and it leads to a tropical paradise located at the center of the earth. It has its own sun, a shallow, warm water sea, and lush, tropical vegetation. What's more the land within is inhabited by a highly advanced and very peaceful race of humans. Normally the hole cannot be seen from the air because of the heavy cloud cover over the North Pole and because the inhabitants of the Hollow Earth keep it covered with electronic 'light screens' These screens give the illusion of vast fields of ice and snow through holographic manipulation of the snow and ice surrounding the hole. Because the hole is so large, the slope down is very gradual and any explorers are hardly aware they are entering another world.

The National Enquirer added by way of a footnote: 'A number of other men claim to have entered the hole at the North Pole. They include William Shavers, a Navy pilot who crashed at the North Pole during World War Two. A tribe of wandering Eskimos also told a Canadian reporter in 1956 that they had found 'a green land at the top of the world.

This last sentence deserves more than just a passing reference because a number of explorers and researchers believe the Inuit are intimately connected with the Hollow Earth legend, and may even have originated there. Certainly, they remain one of the most mysterious people on earth, living in the world's most hostile environment, and about whose origins no one can really be sure. Some authorities believe they are the oldest inhabitants of the northern hemisphere, existing in a region for which nature never intended human beings and into which they came by chance.

The Norwegian explorer Nansen in his book in Northern Mists' (1911), writes at some length about his experiences with the Eskimos or Inuit and declares at one point:

When we remember that in the efforts of the Eskimos to tell us where they came from, they would point to the north and describe a land of perpetual sunshine, it is easy to see that the Norwegians who associate the polar regions with the end of the world, certainly not with a new world, would wonder at the strange origin thus indicated. No wonder we regard them as a supernatural people who may well have come from the interior of the
Earth.

In 1909 while Rear Admiral Robert Peary was exploring at the North Pole, he was surprised to learn that his Eskimo guides believed he was on an expedition to find the "great people" to the north, from which they were descended. Peary understood them to mean a paradise where the inhabitants possessed great powers. From what he was able to deduce of their religion, his Eskimos believed that after death they would 'descend beneath the earth where the sun never sets and the waters never freeze'. Charles Berlitz, in his 'World of Strange Phenomena', tells the story of 2 archaeologists, Magnus Marks and Frolich Rainey, who carried out excavations at Tpiutak in June 1940. There they discovered the ancient ruins of what they could only describe as an 'Arctic Metropolis'. Rows of buried stones and elaborate Eskimo carvings pointed to the fact that there had once been as many as 800 habitations extending along the shore for almost a mile, home to around 4,000 people. According to Rainey this amounted to an incredible number of inhabitants for a hunting village in the Arctic, and he theorized, 'The people of this Arctic Metropolis brought their arts from some center of cultural advance.

In his book 'Not of this World', Peter Kolosimo points out with equal significance: The Eskimos believe they were deported from regions which today are tropical by the use of "huge metal birds" (UFOs yet again!). Another legend among them just as current is that some of their forefathers, now dead or "carried off into the skies", returned afterwards with magical powers they never had before.

A UNIVERSE OF HOLLOW WORLDS?

When Marshall B. Gardner came to write his book 'A Journey to the Earth's Interior' many years ago, he went rather deeper to substantiate the idea.

'When we say that the Earth is a hollow body with polar openings and an interior sun, we back up the statement by referring to nebulas in many stages of evolution in which the gradual forming of the outer envelope of the future planet and the interior sun, and even the beginnings of the polar openings, are all clearly visible in their different stages. Then we point to the actual construction of the planets, Mars, Venus and Mercury, and we show just what the polar openings are like. We show that they are not just ice caps, because direct light has been seen to come from them. And then we demonstrate conclusively that the Earth, like Mars and the other planets, has its polar openings, too.

The idea that the moon could be hollow was proposed in
January, 1970 by 2 Russian scientists, Mikhail and
Alexandra Chtcherbakov of the Soviet Academy of
Sciences. In an article published in 'Komsomolskaya Pravda, the couple rejected the 3 general hypotheses about the Moon 's origin: (1) that it was a piece of the Earth that had been torn away; (2) that it had been formed independently from the same cloud of gas and dust; and (3) that it had wandered into the solar system from far away and been caught by the Earth's gravity.

The Chtcherbakovs said that the Moon's low density compared to that of the Earth suggested to them that it was actually a hollow sphere:

This appears to consist of 2 shells containing an atmosphere, the inner 39 kms thick made of extremely hard metal, the outer about 4 kms thick composed of thermo— protective. resistant and inoxydizable rocks including chromium, titanium and zirconium. The larger lunar craters and their surprisingly small depths were caused by meteorites striking the metallic 'shock absorbers' and exploding sideways not inwards, causing extensive shallow holes and scattering debris far and wide.

What made the 2 Russians' theory really fascinating was their contention that the Moon is actually an 'artificial satellite' that was once launched on a geocentric orbit around the Earth by alien beings of enormous intelligence whose civilization was based on a giant space— ship:

Inside the hollow interior of the Moon are storerooms for propulsive, tools and materials for repairs, navigational equipment and observational instruments. Some lunar rocks are different and older than terrestrial rocks. This does not prove that the Moon was fabricated before Earth was formed, but it is undoubtedly extremely ancient. The satellite, now probably uninhabited, is becoming a wreck, the stabilizers no longer function, the poles are displaced, and the face opposite Earth wobbles badly. The dark 'maria' or 'dried seas' seem to be patches of the metallic inner sphere stripped of its protective sheath, later repaired. The repair material and equipment underneath these localities explain the phenomena of 'mascons, zones of increased gravity discovered in eccentric orbits of our own lunar satellites. Gas sometimes seen escaping through craters is not due to volcanic activity but to leakage of atmosphere inside from fissures in the outer sphere.

Startling as this idea must have seemed at first to many readers, it soon attracted serious attention in America, especially from the science writer Don Wilson, who published a series of essays in 'Fate' magazine giving various examples of how strange a place the Moon quite evidently was. He cited a number of curious phenomena that had been observed and measured on the surface by astronomers all over the world, including clouds of water vapor and mysterious winking lights. He drew attention to the fact that the average density of the Moon was the same as lightweight aluminum, and printed a new photograph of a crater at the Moon's South Pole which was very much deeper than any of the others. He suggested that this just might be an entranceway to the hollow interior, very similar to those in the Arctic and Antarctic regions on Earth, and went on:

The seismic data from the Moon is also extremely impressive. Scientists were literally falling off their chairs when it first started to come in. The analysis done by the scientists at NASA in 1962 seemed to point to only one conclusion the Moon must be hollow!

In astronomy, mass is dealt with on the basis of relative masses. This being so, if the Moon is hollow, then so might be the Earth and other planets and their satellites because gravity is based on masses relative masses and distances. Therefore, if only one object was hollow it would be patently different from all the rest.

Subsequent study by a number of Hollow Earth researchers has produced a list of those planets and moons most likely to be hollow; and I propose to look at the evidence for them one at a time. They are, after the Earth and the Moon: Mercury, Venus, Mars and its moon Phobos, Jupiter and its satellite Callisto, Saturn and, just possibly, the 2 very recently discovered moons of Uranus.

MERCURY is, of course, the planet nearest the Sun and is believed to have the harshest environment of any world in the solar system. Its appearance is not unlike that of the Moon, with many well—defined craters and, significantly, an interior said to 'very strongly resemble our own ', according to a NASA scientist after the fly past of Mariner 10 in 1974. Mercury has a curiously weak magnetic field which gives it an unusual axis of spin and this is what first led to suggestions that it might be hollow.

Astronomers in the past have several times noticed a tiny, bright dot on the planet during a Mercurial eclipse. This caused Raymond Palmer to speculate, in an article in The Hidden World, 'if it isn't true that just at that moment, the polar orifice of Mercury, formed as other bodies by a vortexial action, presents itself at precisely the proper position to be observed?' Further discussion has followed a report in the April 1992 issue of 'Final Frontier' on this same phenomenon. 'Despite temperatures that can climb as high as 800 degrees F, ' the journal stated, researchers stated, ' researchers at the California institute of Technology in Pasadena have identified what they believe is a water ice—cap more than 180 miles in diameter on Mercury's north pole.

VENUS is often referred to as 'Earth's Sister Planet ' being our closest neighbor at a distance of just 26,000,000 miles. Space probes from both Russia and the USA have penetrated the blankets of yellowish clouds which cover the planet. These proved to be almost 95% carbon dioxide with a hint of Sulphur dioxide providing the color while traces of water vapor were also recorded. The surface of Venus is covered by vast rolling plains and 4 major highland regions dotted with a number of active volcanoes.

The annals of astronomy tell us that twice, in 1686 and 1833, bright lights were observed shining out of Venus' North and South Poles.

Then, in 1978, a Pioneer 12 orbiting vehicle revealed t holes' in the atmosphere just above the North and South Poles. A decade later even more significant evidence was revealed when NASA's Jet Propulsion Laboratory released a series of radar—generated photographs of the planet which pierced through the clouds, one of the images clearly showing a north polar opening. The picture had such an impact on the scientific community that it was published on the front cover of the April 1989 issue of 'Discovery' magazine.

MARS, the 'Red Planet', has probably excited more interest than any of the other worlds in the solar system and, certainly, the most widespread speculations that it could contain life. The stories about its so—called 'canals' are part of scientific folklore, and books such as H. G. Wells' The War of the Worlds (1898) have given it a unique place in speculative fiction. With its erratic motion, white— capped poles and dark, blue—grey formations in a constant state of change, the chances of it being hollow are considered to be very high. Mars has a day almost identical to ours, although its atmosphere is much thinner and consists mostly of carbon dioxide. The temperatures are such as to prevent water existing for long as a liquid, changing directly from vapor to ice and back again.

The US astronomer Percival Lowell (1855—1916), who first suggested there might be life on Mars as a result of observing the criss—crossing pattern of lines he identified as 'canals', was also the first to theorize that the planet might be hollow. He spotted through his telescope 'a dark, circular band' appearing around the poles during one summer season, and the following spring noticed something even stranger. 'As I was watching the planet, I saw suddenly two points like stars flash out in the midst

of the north polar cap. Dazzlingly white upon the duller white background of the snow, these stars shone brightly for a few moments and then s lowly disappeared. Could these have been rays of light from a central sun inside Mars? A contemporary science writer, Martin Caidin, has also reported that in recent years several American and Russian astronomers have observed very bright flashes originating from both the North and South Poles of Mars. Ernest L. Norman, an American scientific researcher of the paranormal claims in his book 'The Truth about MARS' (1998) that not only is the planet hollow, but that it is actually inhabited by a highly intelligent humanoid race.

He writes: On Mars the cities are all underground, and as the outside temperature is very rare and of a low oxygen content, they are becoming less and less dependent on that source of air supply. Many thousands of years ago the Martians learned how to obtain air from water by electrolysis... They are a quiet, peace—loving people who originally migrated there in a space craft from a dying planet and settled below the hostile surface. . .

PHOBOS is the closer to the planet of Mars' two moons about 5,830 miles away and a strong case has been made that it, too, is hollow inside. The little satellite, about 23 miles in diameter, has a unique orbit of just over 7 hours, which means that it travels round the planet twice in a Martian day. The moon's surface contains numerous grooves up to 220 yards wide and several craters, the largest of which has been named Stickney. All are believed to have resulted from the impact of missiles from space.

Photographs of Phobos taken by the Viking spacecraft in the late Seventies give it the eerie appearance of an alien craft in perpetual orbit around Mars.

JUPITER is the largest planet in the solar system — 1,300 Earths would fit comfortably into its bulk and the first of the four planets, beyond the asteroid belt, which make up the Jovian group with Saturn, Uranus and Neptune. It is famous for its 'Great Red Spot, a strange phenomenon big enough to hold 2 Earths, and is a mysterious world that has fascinated astronomers for generations. There was general amazement, however, in 1979 when Voyager 1 flew past Jupiter, revealing that the spot is actually a vortex piercing the planet 's three distinct zones of water, ammonium hydrosulphide and ammonia, right down to the surface! This has prompted speculation that Jupiter could be in fact hollow, with at least one entranceway far more visible than those on Earth or any of the other planets. Voyager 1 also confirmed that Jupiter was probably originally formed from a swirling mass of dust and gas and can actually be compared to a gigantic 'bag' of gases, consisting of hydrogen and helium in vast 'shells' thousands of miles thick. Just what lies at their heart is the real puzzle of Jupiter.

CALLISTO, the moon of Jupiter which has the great possibility of being hollow, along with many others, is said by NASA scientists to be per— haps the only spot in the Jovian system upon which mankind might feasibly land at some time in the future. The farthest of Jupiter's 4 major Galilean satellites, named after Galileo who discovered them in 1610 (the other 3 are 10, Europa and Ganymede),
Callisto is about 3,000 miles in diameter and is believed to be covered by a soft ice mantle and a thick ice crust. It has probably more craters on its surface than any other moon or planet in the solar system, the result of being regularly bombarded by meteors throughout its entire existence. The most curious sight of all is a 378 mile—wide crater at the heart of a series of concentric rings extending for at least 620 miles in all directions. Could this be an opening to the interior?

GANYMEDE, 2 photos of which are in my volume 1 of 'GRAVITY r MATTER & SPACE TRAVEL' has a very advanced civilization of humanoids living there and their spacecrafts have visited our planet for a very long time. Ganymede is twice the size of planet Earth and unique, as can be seen from the unique 'features displayed by the photos of this very fascinating world. This advanced civilization has been there for at least 7,000 years.

SATURN is one of the wonders of the solar system with its unmistakable ring. It hangs in the heavens like a huge, glowing ball with its cluster of satellites, orbiting leisurely around the sun once every 29 years. For a long time, Saturn was mistakenly believed to be 3 planets, and it was not until 1659 that a Dutch astronomer,

Christian Huygens, defined it as a single world 'surrounded by a thin, flat ring, which nowhere touches the body'. Tt was Voyager 1 which provided the first close—up views of this romantic world, revealing that the ring—system actually consists of 300 'ringlets' of solid particles of rock and ice. These do not gradually merge into one another as conventional theory would have it, but are separated by complicated gravitational interactions between the ring material and 6 of Saturn's 15 moons. Whether the rings were created at the same time as the planet, or are the remnants of a comet or moon that passed too close and was torn apart and swirled into orbit, remains a mystery. Saturn takes about ten Earth hours to revolve, and is flatter at the poles than most other planets in the solar system. The complexity of the ring has generated much speculation about the true nature of the planet, not forgetting the white oval spots on the surface which the NASA space— craft also photographed. The bands of lighter and darker clouds, indicating equatorial and temperate belts, are equally puzzling, although once again, the fact that the planet consists of ice, hydrogen and helium with a certain amount of liquefied rock has fueled speculation that it, too, may be hollow.

URANUS, the blue—green disc on the far frontier of the solar system, is the third— largest body in the system. With a diameter 4 times that of Earth, it is not unlike Jupiter and Saturn and possesses a dense hydrogen atmosphere. Uranus has an extraordinary orbit, with an axial tilt of 98 degrees relative to the plane of its orbit — which means that first one pole and then the other points directly at the Sun during the course of its 84— year orbit. Both of these poles are noticeably flattened. In complete contrast, the planet rotates very quickly on its axis, turning once every 15 hours. Naturally, this results in the most bizarre 'seasons' on Uranus, with long, intense periods of light and dark. A number of dark spots have been sighted on the surface along with 2 rather dusky— looking poles, and these have prompted speculation that there may well be disturbances taking place inside the planet.

Two small newly identified moons, both far distant from Uranus and as yet unnamed, have added to the planet's mystery. They were discovered in April 1998 by a team of astronomers led by Dr. Brett Gladman of the University of Toronto, using cameras mounted on the 16—foot Hale Tele— scope at Palomar Observatory. The pair have been added to the planet 's known 5 Oberon, Titania, Ariel, Umbriel and Miranda and are said to be 40 and 80 miles in diameter respectively and unusually red in color. Initial observation of the moons by Dr. Gladman suggests they may have originated in the outer solar system and been 'captured' by the gravitational pull of Uranus. Again, the similarity between these two curious bodies and Phobos and Callisto has led to speculation that they might also be hollow.

UFOs FROM THE HOLLOW EARTH - THE ANTARCTIC

In this chapter we shall be examining evidence from the nations closest to the Antarctic, dating back many years, starting in 1912 over New Zealand.

In 1912, a resident of Dunedin, some miles up the east coast from in— vercargill, saw a mystery object. An account of this, by Isabella Walmsley of Christchurch, was quoted by Harold T. Wilkins in his book 'Flying Saucers on the Moon (1954). Mrs. Walmsley also referred to an experience of her own, some years later.

Everybody, of course, thought this man in Dunedin was a bit 'touched ' when he claimed to have seen a strange object in the sky going south along the coast. He must have seen a meteor, folks said... Now, I am a light sleeper, a legacy from my days as a nurse, and one night, when we were living in Timaru (also along the coast about 100

miles from Dunedin), I was awakened, suddenly, by a loud, roaring, hissing noise that passed swiftly over the house. It was the year 1935, before there were night— flying planes here. When the Second World War broke out, T thought it must have been a Jap plane, but it did not occur to me to wonder where the base of such a rover could be, with thousands of miles of open sea all around New Zealand! A few days later, I was talking to an old chap who lived in the neighborhood and he told me that he had to get up in the night, and when he got outside into the open air he saw a 'big light like the sun, moving south over your house, missile! In April 1952, I spent a week in Dunedin and was similarly wakened by a loud swish overhead going south. It was still, and about 4 a.m. There was no fading. The noise was there, and then it wasn't! I did not mention it to the people I was living with as T knew how they would regard the story. a few days later, the New Zealand and Australian newspapers were full of flying saucer stories.
The year 1960 brought another curious report which fits into more UFOs on the move.

On 3 October, six flying saucers and a "mother ship" were reported over Tasmania. A Church of England minister on the island said he had first seen the mysterious craft nearly a week earlier, but was reluctant to report them until other people in the area told the same story. All agreed that the objects appeared and disappeared to the south.

A similar account of a mother ship and small objects over the coast of South Australia in 1961 is reported in detail by Jacques Val lee in his 'Anatomy of a Phenomenon' (1965). A group of round, silvery objects was observed leaving and returning to a bigger object which finally collected them all and departed at high speed southwards over the Great Australian Bight.

Australia had another bumper year of sightings in 1965, starting on 25 July when a number of children in Melbourne reported a bright, silvery object that maneuvered in the sky above them, rose vertically, gave off bright flashes, and then disappeared out to sea. A metallic disc was also observed racing to the southern horizon from Australia's farthest —flung territory, Macquarie Island, in November.

By far the most extraordinary story to come from
Australia, however, occurred over the Bass Strait between Melbourne and Tasmania on 21 October, 1978, when a pilot was lost in what is alleged to have been a UFO abduction. In the early evening, 20—year— 01d Frederick Valentich set out in his single—engined Cessna 182 on a short solo flight from Melbourne to King Island. He was about 95 miles southwest of Melbourne when he reported to ground control on the mainland that he was being followed by 'a high— speed object with green lights'. Asked by a controller to identify the craft as there was no other traffic in the area, Valentich insisted it was not an aircraft. After a brief break in transmission, the pilot came back on the air again: 'It's coming for me now. I'm orbiting (circling) and the thing is orbiting on top of me also...it has a green light and sort of metallic light on the outside. ' At this, a long metallic noise was heard and Valentich and his plane vanished without a trace.

2 other light aircraft have also disappeared in mysterious circumstances while flying over the Bass Strait. In December 1969, a Fuji single—engined aircraft with only the pilot on board was lost on a flight to King Island; and in September 1972, a Tiger Moth with 2 people on board vanished on a similar trip from the

mainland. After extensive and fruitless searches, the official explanation was that all three pilots had probably confused the lights with 'the planet Venus which is very bright at this time. The statement was greeted with understandable skepticism, even ridicule. What no one was able to explain was that in each case the UFOs had first been seen materializing from due south.

In the spring of 1962 it was the turn of the Argentine navy to see strange objects in the sky. According to the Buenos Aires newspaper El Mundo, four sightings were made at sea two by naval pilots and two by ships' officers. later the same paper reported that on the night of 15 June bright objects flying high in the sky', while at the neighboring coastal resort of Miramar a cigar— shaped object was seen at 9:30 p.m. flying in from the direction of the South Atlantic. The whole craft was vividly illuminated. El Mundo said, and carried 3 very bright lights red in the center, yellow on the right, and green on the left. 'The flood of these cosmic sightings of late has had a profound emotional impact on the many eyewitnesses, ' the paper added, 'and may well mark a new era with regard to the UFO problem.

2 subsequent events in Argentina have also earned considerable media attention and comment. On 17 July, 1965, a circular object which reflected the rays of the sun on its metallic and polished surface' hovered over a beach near the River Plata, where it was seen by hundreds of bathers and prompted a demand from La Nacion the following day: 'We do not believe the true explanations of such occurrences can be kept secret much longer.

THE PRIVATE DIARY AND LOG OF ADMIRAL BYRD

The following pages are taken directly from excerpts of the explorer, Rear Admiral Richard Byrd. It was obtained from the Hollow Earth Society' of Australia.

It certainly seems valid that his wish to see the land beyond the pole DID COME TRUE! The document is a lengthy one which has clearly been annotated, but the extracts included below do seem to bear the style of Richard Byrd's writing and manner of ex— press ion. IT
IS NOW FOR THE READER TO JUDGE WHETHER THEY

REALLY COULD BE GENUINE. . . LET THE FOLLOWING

FACTS SPEAK FOR THEMSELVES, READ ON. FLIGHT LOG - CAMP ARCTIC FEBRUARY 19th 1947 Vast ice and snow below. Note coloration of a yellowish nature. It is dispersed in a linear pattern. Altering course for a better, examination of this color pattern below. A reddish— purple color also. Circle this area two full turns and return to the sine compass heading. Position check made again with base. Relaying information concerning coloration in ice and snow below.

Both magnetic and gyro compasses beginning to gyrate and wobble. We are unable to hold our heading by instrumentation. Take bearing with sun compass but all seems well. The controls are seemingly slow to respond, have a sluggish quality. Yet there is no indication of icing!

In the distance is what appears to be mountains. 29 minutes of elapsed flight time and first sighting of mountains; it is no illusion. They are mountains consisting of a small range I have never seen before.

Altitude change to 2950 feet.

Encountering strong turbulence again. We are crossing over the small mountain range still proceeding northward as best as can be ascertained. Beyond the mountain range is what appears to be a small river. A valley with a small river running through the central portion. There should be no green valley here. Something is definitely wrong and abnormal here.

We should be over ice and snow. From the port side there are great forests growing on the mountain side. The instruments are still spinning. The gyroscope is oscillating back and forth.

I alter the altitude to 1400 feet and execute a sharp left turn to better examine the valley below. It is green with either moss or a tight—knit type of grass. The light here seems different. I cannot see the sun anymore...

We make another left turn and spot what seems to be a large animal of some kind below. Appears to be an elephant. No, it looks more like a mammoth—like animal, this is incredible, but there it is. Decrease altitude to 1000 feet and take binoculars to better examine the animal definitely a mammoth— like animal. Report this to base camp.

Encountering more rolling green hills. The external temperature indicator reads 749 F. Continue on our heading. Navigation instruments seem normal now. I'm puzzled over their actions. Radio is not functioning. Ahead we spot what seem like habitations. This is impossible! Aircraft seems light and oddly buoyant. The controls refuse to respond.

I tugged at the controls again. They will not respond. The engines of our craft have stopped running. The landing process is beginning. The downward motion is negligible and we touch down with only a slight jolt. I'm making a hasty last entry in the flight log. I do not know what is going to happen now.

Some 'hollow earth' researchers who have studied the log believe it does indeed record a journey 'beyond the Pole' and in through the polar opening to the hollow earth of mountains, lakes, rivers, vegetation and life forms as legend has maintained, including this author! After landing, Byrd met a group of tall, blond men who spoke English in a 'slightly Nordic or German accent' and greeted him with the words 'Welcome Admiral to our domain — you are in safe hands. After this he travel led in a strange disc— shaped machine 'with markings like a type of swastika', and was given a conducted tour of the inner world, visiting a 'glowing city which appeared to be made of crystal'. Later Byrd was allowed back to his aircraft and told to take a message to all surface dwellers about the dangers they faced if they continued to experiment with atomic weapons and other nuclear power stations!

There are many researchers who are and were aware of the stories of Nazis and secret polar bases, this incident with Admiral Richard Byrd is a true account The 'lake Vostok' area in the Antarctic, presently being explored by
American, German and Russian scientists seems to be +03 of primary interest due to its magnetic anomalies. Further studies need to be conducted to determine if it is a possible 'south polar' opening location?? Hopefully, we may have some answers by 2012. In 1955 Byrd was again 'under orders' to return to the South Pole and continue mapping the territory. Byrd told reporters, 'This is the most important expedition in the history of the world.' He made several other flights, with a total of 437,000 square miles of Antarctic territory being discovered and mapped.

This communique completely ignores a much more revealing radio announcement issued from Little America on 14 January, 1956: On January 13, members of the U.S. expedition accomplished a flight of 2,700 miles from the base at McMurdo Sound, which is 400 miles west of the South Pole, and penetrated a land extent of 2,300 miles beyond the pole. A quick look at a map will immediately reveal the discrepancy in this report. The Antarctic is surrounded by water, and it is impossible to travel a distance of 2,300 miles in any direction without traversing water. So where could this land over 2,300 miles away be but inside the Hollow Earth?

In public at least, Byrd was circumspect when he returned from the South Pole to be greeted as the world's greatest explorer. The expedition had, he said, 'opened up a vast new territory' and he spoke lyrically of 'that enchanted continent in the sky, land of everlasting mystery'

If the Rear Admiral had discovered an inner world which could change forever mankind's perception of the earth, the media, it seemed, was too preoccupied to take any more notice. Or was it? Raymond Bernard, in his pamphlet 'The Hollow Earth' (1969), thinks the explanation is rather different:

_____

Admiral Byrd's discovery is today a leading international top secret, and it has been so since it was first made in 1947. After Byrd made his radio announcement from his plane after the brief distress notice, all subsequent news on the subject was carefully suppressed by the government agencies. The explanation is evident. If Admiral Byrd made such a very Aomentous discovery, undoubtedly the greatest one in history, of a new unknown land area of undetermined extent, over which his expedition flew for a total of 4,000 miles at the
NORTH POLE AND THE SOUTH POLE, and which area is probably as wide as it is long, and, since Byrd I'm certain that he was the victim of a conspiracy theory $ Rear Admiral Richard E. Byrd was silenced. When he died in Boston on 11 March, 1957, the authorities honored him as a national hero and he was buried in Arlington National Cemetery, Virginia, with full military honors. But to the grave with him went the secrets of precisely 'WHAT they had seen while flying over both poles. A final paragraph in his diary does not answer all the questions, but certainly indicates that Byrd had believed there was more to the Arctic and Antarctic wastes than just endless miles of frozen snow:

December 24, 1956:

These last few years since 1947 have not been so kind. I now make my final entry in this singular diary. In closing I must state that I have faithfully kept this matter secret as directed all these years. It has been completely against my values and moral rights. Now I seem to sense a long night coming on, and this secret will not die with me, but as truth shall, it shall triumph. It is the only hope for all man— kind. I have seen and it has quickened my spirit and set me free. I have done my duty towards the monstrous military industrial complex. Now the long nights of the Arctic and Antarctic end, the brilliant sunshine of truth shall come again, and those who are of darkness shall fail in flight. FOR 1 HAVE SEEN THAT LAND BEYOND THE POLES, THE CENTER OF THE GREAT UNKNOWN.

Underneath is scrawled the familiar signature, 'Richard E. Byrd, United States Navy.

THIS SCIENTIST IS DEDICATED TO CONTINUE WORK ON
HIS "MAGNETIC MOTOR" AND GENERATOR FOR FUTURE
DEPLOYMENT BY 2012-2013. YOU CAN BROWSE OUR WEBSITE @   www.mullerpower.com dedicated to the late Bill Muller

of Penticton, B.C., Canada. You may contact this author @

(760) 327-4761

HOME PHONE ONLY FOR ANY QUESTIONS YOU MAY HAVE AT YOUR CONVENIENCE.

MAY GOD BLESS YOU ALL, Hans J. Petermann, SCIENTIST
+ ELECTRONIC ENGINEER

INCREDIBLE ENCOUNTER WITH A FEEDING UFO OR

_____

SOMETHING FAR MORE DANGEROUS

This last part is a transcription of a chapter taken directly from the book "STALKING THE TRICKSTERS" by Christopher O'Brien, published in 2009 by 'Adventures Unlimited Press, Kempton, Illinois.

It was years ago, about 1995. I was out setting up for a big hunt. It was me and 2 buddies of mine we went out about 2 in the morning to drop off traps and other things we were going to use during the day's hunting trip out near the end of Long Island.

My one buddy was a retired phone man and knew the right of way used by the utility companies and had a good sense of the woods in the area where we were dumping our supplies for the day. The three of us packed up the gear we needed to carry out to our meeting point later that day and started our trek following the right of way into the woods (toward some transmission towers). We kept walking towards the center of the grouping of electric transmission towers until we could see clear of the trees (a power blackout had just occurred) The 3 of us saw it all at once... we could not believe what we were seeing. Ahead of us over the group of towers, we could see the laser light was coming straight down from the sky. We followed the light up and could clearly see a huge black round saucer shaped craft hovering silently over the towers. The thin light attached to the wires and pole. There was no sound, no movement, just this light drinking or doing something with the towers and wires. The thing was huge as when you looked up at it the entire sky was blocked out but for the massive shape of the craft. Around the borders of the craft you could see the night with all its stars and moonshine.

I called the 3 smartest people I have known for advice on this encounter... I was fascinated and stunned to be told by all 3 the same thing! None of the men thought this craft was feeding at all. In fact, they all felt that this craft was putting in data not taking anything out! All of them also added in a matter of fact way that "of course they were also taking all the data that was being transmitted out through the lines as well."

I was told that it most likely was a case of catching the UFO placing in data that could and would be easily transmitted into the lines, cables, and communication systems found at that location. The question is what type of data were they feeding into our lives?

My one friend the physicist told me he thought it possible that we were being fed messages of some type of control or illusion via the wires into our televisions, computers, even phones and radios that are in every home in America, and in fact, most homes on the entire world's surface.

My one friend added as we parted, 'I think you walked into something very big with this one dear. I think those men walked into one of the biggest stories going and one that has most likely been going on since we primitive beings invented the wires that hook into each and every home, office and building on earth!"

So what was the description of the craft's actions in relation to the area's power blackout? If these smart folks are correct, and the object was inserting something into our power and communications grid: what data is being programmed into our infra— structure and by whom? I can think of no better term to describe the engineers of this event than "tricksterish" regardless of their agenda...

I then asked her what she thought about power blackouts, UFOs and their possible correlation:

Blackouts and UFO sightings are common and have been since we beings started placing uniting wires between our shelters. The 1964 New York City blackout came with thousands of reports of UFO sightings, lost time events and claims of abductions all along the northeast coast, T was told by one electrical engineer and one group of electrical utility workers that my article gave them clues to what may have been the cause of many malfunctions and area blackouts over the years that they had worked on that had no explanations and corrected without anything being fixed!

Whether the skeptics and debunkers like it or not, UFOs have become monstrous 'super—gadgets' in world culture. The subject fascinates millions around the world and shows no sign of releasing its grip on the popular imagination anytime soon. There are many, many areas that I could dive into regarding this subject, e.g., man— made UFOs, is— closure by the government, abductions, secret military technology disguised as "UFOs," attendant phenomena, to the real thing etc., as I had previously mentioned in my book. I'd like to point out a few salient points that are often overlooked by the true believers in 'ET' crowds worldwide.

First of all, when it comes to reports of witnesses seeing what appear to be UFO occupants, there has been a definite progression over time. The same holds true with descriptions of their craft. Stretching back to biblical times, these aerial objects were described as flaming chariots, wheels in the sky, flying shields, flying seed pods, etc., in the late 1890s they appeared as steam—powered diri— gibles; today they are more high—tech than we can imagine. Obviously, witnesses revert to familiar images when attempting to describe some— thing outside of their normal frame of reference, but these descriptions down through history are fairly obvious they are describing what we now call UFOs. However, I find it curious that today's most enduring image of the pilots of these craft is that of 'almond— eyed' diminutive gray aliens with oversized heads. This description can now be found in witnesses' accounts from all around the world. But this wasn't always so. Prior to the release of Whitley Strieber's influential book, 'COMMUNION' with its iconic cover image of an almond—eyed gray, little grays were never seen outside of the USA. South of the border, in South American and the Latin countries, they were sometimes described as dark, hairy dwarves. In European countries they were most often human—appearing, and the further east you went into Russia

they even have been reported as having robotic forms, as in the infamous 1989 Voronezh sighting wave.

Because of the media coverage, and the resulting pop culture perception, the "grays" now have been divided into several subgroups and types, and our cultural view of this particular type of being appears to be evolving symbiotically right along with the various guises of the aliens. I've often wondered, why are these beings constantly observed examining rocks and picking flowers as if they have never seen them before?

If they had even our existing level of technology, they could put an object the size of a beer keg in orbit around the Earth and find out whatever they wanted to know about our ecosystem; but no the aliens still land and walk around like they have never been here before. And what's up with the medieval medical experiments and cattle mutilations worldwide? Some descriptions by abductees refer to experiments that seem downright antiquated by our current technological standards. I thought they were a million years ahead of us. Basically, what I am driving at is that there appears to be something besides "extraterrestrials" flying these craft and supposedly ab— ducting folks. Jacques Val lee nailed it 40 years ago in 'PASSPORT TO MAGONIA ' — it's a new, high—tech face on an age—old mystery. I have been asked countless times what I think about "the aliens". For years my stock answer has been: "I don't believe they 're aliens; T actually think 'WE ARE THE ALIENS. ' They are probably more terrestrial than we are, as they have co— existed with the dinosaurs and set up under— ground bases at least 100,000,000 years ago in their caves and inter— connected tunnel systems worldwide! Furthermore, pertaining to the 'Roswell' incident in New Mexico back in 1947, my late father who retired as Colonel in world war

2, told me that while he was working in Peenemunde in Germany from 1940 - 1945 that the Germans had developed their own 'flying manta—ray type discs using advanced plasma fusion' technologies! THIS IS AN ABSOLUTE FACT unfortunately, the entire incident was covered up by the US government The 'Kecksburg', Pennsylvania incident in 1967 was also another project that crashed into the woods there, project being under control of the 'US military' covert operations! Anyway, the 'black projects' at areas S4 + 51 in the Nevada desert also involve very exotic 'back— engineering of 'BIOPLASMIC FUSION' technologies, dear readers! Those 'Phoenix, Arizona' lights from 1997 were manufactured, courtesy of the S government... There are trillions of $$$ that have been spent on these 'black projects' for many years! NO WONDER THE USA HAS BEEN BANKRUPTED FOR MANY YEARS DUE TO THE CORRUPTION REIGNING RAMPANT IN WAGING WARS ELSEWHERE, AS "PRESIDENT OBAMA" IS PART OF THIS MONSTROUS MILITARY INDUSTRIAL COMPLEX, AS THESE WARS WILL NEVER END BY 2014!

Finally, photo is enclosed showing a dead gray alien being. The late Phil Schneider should be highly recommended for his bravery in really exposing the monstrosities of this totally corrupt system!

QUALITY EDUCATION SHOULD BE THE TOP PRIORITY FOR OUR CHILDREN, INSTEAD OF THE POOR EDUCATIONAL.

The photograph of the ET "Diplomat" shows the upper torso, shoulder, neck and head of the ET life form looking directly into the camera. One can barely make out two over—sized, darkened eyes, heavy brows and the back of the head is just barely discernable. According to Dr. Greer this ET was between 3 and 5 feet tall and 'hovered about 2 to 3 feet above the ground. ' It is certainly the first publicized photo of a life form taken during one of the 'Ambassador to the Universe' training sessions. This may be the first, or one of the first, confirmed photographs of an ET being at close range. The greenish coloration is apparently the ET being's uniform. The source of the bright reddish coloration is unknown. It could be the 'robot 's bioplasmic' system of operations', 'TERMINATOR' style. The bright white and blue image to the left of the photograph is the armrest and back of a camp chair. The photograph was taken without a flash attachment and was hand— held.

Source: www.disclosure project Photograph of ET
"Diplomat"

Rear Admiral Richard E. Byrd
whose polar explorations led him
to fly into the Hollow Earth.
(Below) Open water in the midst
of the Antarctic photographed by
Rear Admiral Byrd's navigator,
Captain McKinley.

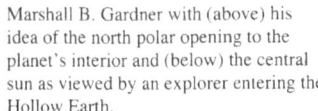

Marshall B. Gardner with (above) his
idea of the north polar opening to the
planet's interior and (below) the central
sun as viewed by an explorer entering the
Hollow Earth.

MORE ANTARCTIC SECRETS BELOW FROZEN CRUST!

It's winter now in Antarctica and very cold with temperatures going down to — 100 degrees F. Given such conditions, how can this utterly cold, desolate, and remote place sustain life in operations down in this amazingly huge continent?

Valmar Kurol of the Montreal Antarctic Society/ Societe Antarctique de Montreal quotes polar explorer Rear Admiral Richard E. Byrd, in his October 1947 'National Geographic Magazine' interview, as follows:

"There is no other music like the tone—less music of millions of years of accumulated silence, through which come bars of unearthly colors.

There is no need for ears to hear the fugues played on this ice organ."

The eloquence of these 2 sentences says it all on the musical and poetic side, but we are merely warming to our topic. Will our delving let us strip away not just   the thousands of feet of Antarctic ice, but also, perhaps, the straitjacket of which our   preconceived notions about the place are composed? Let's begin with the Piri Reis map. INCREDIBLE MAP MEETS AMAZING THEORY

Cartographers have long been baffled by the astounding map of Turkish Admiral pi ri Reis, a map known to be based on reputable sources dating back to at least 1, 000 years ago. Why astounding? Because it shows what Antarctica looks like without the ice a condition which has not occurred for thousands of years! Such knowledge has, in theory, only become available as a result of modern satellites and sounding techniques, in turn leading to speculations about ancient astronauts from elsewhere as the only 'rational' sources for such information. Suppose, though, that a better, simpler explanation were available, an explanation which both fully satisfied Occam's Razor and made myth concrete?
What then?

CRUSTAL SHIFT, BUTTERCUPS AND ATLANTIS

The 1513 Piri Reis map has made things tough for cartographers, but paleontologists have been

more sorely beset trying to explain, typically using a geological model based on gradual change, how Siberian mammoths came to be flash frozen so quickly that their last meal, of buttercup flowers, lay undigested in their stomachs. Call some of their stories a stretch! What isn't a stretch, though, is what could have happened under the crustal displacement theory of Professor Charles Hapgood who, in his 'Earth's Shifting Crust: A Key to Some Basic Problems of Earth Science, Basic

Books, 1958', argued that a cumulative imbalance of ice and snow could cause the earth's crust atop the fluid mantle to suddenly shift, turning whole regions into instant deep freezes. A stray comet could have totally destroyed the islands of 'Atlantis', in addition to the destructive powers of these very destructive 'super— crystals' that were being used, as mentioned previously!

HITLER'S ASHES, UFOs, AND SECRET WARFARE
PART 2

Despite a popular perception of Antarctica as being a frozen waste— land, there exists a largely hidden history of a different Antarctica, of a habitable thanks to geothermal soil heating zone in Antarctica, of a NAZI— CONTROLLED piece of Antarctica, as I had mentioned in my previous book entitled "GRAVITY, MATTER & SPACE TRAVEL", PUBLISHED IN 2006, by Trafford Publishing. Please study the relevant chapters in that book pertaining to the 'South Polar' opening of 140 miles in diameter at   'NEUSCHWABENLAND'. Please view the back cover of it, indicating the various tunnels and regions...

Look for the region called 'Queen Maud Land' or Dronning Maud Land. By meticulously photographically surveying the region and dropping weighted marker posts topped by 'Hakenkreuze ' swastikas to delineate their claim and magically capture it. This was all accomplished back in 1937 1940 under Captain Ritscher and various submarines
& ships!

Neuschwabenland became and was treated as part of the Third Reich! Moreover, it became the site of very intense covert scientific research and colonization, under the name Base 211, starting around 1941. The base vas situated in the aforementioned ice—free areas and reportedly also enjoyed underwater access from the sea via ancient volcanic tunnels and man— made tunnels, making it perfect for U—Boats. What's also remarkable about this area is that a range of many sources explicitly link it to German flying saucers. These sources include:

W. A. Harbinson's massively researched and foot—noted novel 'GENESIS' information from German secret societies, and Top Secret SS blueprints in the video "UFO SECRETS OF WW2 GERMAN FLYING SAUCERS, " excerpts from Rear Admiral Byrd's purported diaries, personally confirmed years ago by Byrd's nephew, Harley Byrd. Furthermore, I was at the "Tesla Symposium" in Colorado
Springs, Colorado as a Science lecturer from the years
1993—94, 1995— 1997. At that symposium I met Al
Bielek during these 5 years. He confirmed to me that
Hitler's flying saucer programs definitely bankrupted
Germany...I found out from my late father that the secret
German organization 'ODESSA' translated to
'ORGANTSATION DER EHEMALIGEN SS ANGEHbRIGEN' - had moved all of their very precious metals and equipment into the countries of Paraguay, Argentina and Brazil! In a series of shocking interviews given in 1947 by then Rear Admiral Byrd, warned people worldwide and to several South American newspapers in particular, that there was and still is a very serious "threat

from both poles, underwater, in the ionosphere and elsewhere on planet Earth. Such candor came to a screeching halt when he got back to the USA.

What prompted such actions? He lost a bunch of naval C—47s equipped for photographic survey.
They were taken down on the aircraft carrier U.S.S.
Philippine Sea as part of the 13—ship 'Operation
High jump' military expedition and flown to 'Little
America ' the U.S. Antarctic base. He was shown by German— speaking saucer crews in face—to— face meetings, in no uncertain terms, just who had the real power! The hidden histories indicate that he didn't take this lying down, then returning a decade later, under cover of an International Geophysical Year expedition, to deliver 2 nuclear weapons that resulted in not wiping out Base 211. These actions are alleged to be the real reason for the hole in the ozone layer!

As if this weren't enough to consider, we must also take into account Grand Admiral Karl Doenitz's claim that the Kriegsmarine (German navy) had built an ' impregnable fortress for the F dhrer' in a faraway location and the statements made to author and WW2 veteran Howard A. Buechner by a former SS officer who said that he was on the U—977, known to have finally surrendered in Argentina 11 months after the war ended, along with U—530 and that they went to Antarctica on a vital Nazi errand. What errand? As related in 'Adolph Hitler and the Secrets of the Holy Lance (Thunderbird Press, 1988) and in Hitler's Ashes Seeds of a New Reich' (Thunderbird Press, 1989), the cargo was indeed Hitler's ashes, the notorious Blutfahne — blood flag of the Munich beerhall putsch, seen in the Ndrnberg Rally footage being used by Hitler to magically pass its essence to and 'bless' through the Law of Contagion the flags he touched with it, and the real Spear of Destiny AKA Lance of Christ. The one found by the Allies was reputedly an expert forgery by a Japanese master swordsmith.   In occult and practical political terms, he thus reportedly carried with him not only the core symbols of Nazi ism, but the ritual tool by which to resurrect it and give it real effect— the Spear of Destiny. Since neither book so much as mentions a base, why use Antarctica as secure storage if there wasn't one? Even more bizarrely, if we accept that the Nazi core was stored in an ice cave, how was it found in 1977 when the Blutfahne and the Spear of Destiny were supposedly slipped back into Europe into the 'Wewelsburg' castle in Germany after a secret light aircraft visit to Antarctica by the same former SS officer? This objection makes eminent sense considering Richard Byrd was barely able to locate Little America only a few years after leaving it because everything but part of the smokestacks and the radio tower had been buried by the snow. The former SS officer asserts that the Spear was to be used to transform Europe. Did this alleged operation have anything to do with the fall of the Soviet Union in 1989 and the reunification of Germany, or the rise of hedonism movements in Europe? If the timeline is correct and a dummy was left for the Allies to find, then there was plenty of time to replace the dummy with the real one before the recent forensic investigations were finished.
'

The Lake Vostok Mystery & Anomaly

These files about atomic bombing in the Antarctica

http://www.thelivingmoon.com/47brotherthebig/03files/_Part_003.html confirmed to Thomas Greanias by

National Science Foundation personnel, as stated in 'The Secrets of Atlantis Revealed:

Frequently Asked Questions.

Richard Hoagland's Enterprise Mission www.enterprisemission.com noticed the apparent existence of some sort of medical crisis among U.S. personnel in the American Antarctica base in the work. According to several articles there, the numbers of bizarre, urgent med evac cases among the care— fully screened staff suggest biological and/or radiation exposure.

Weirdly, this is apparently backed up by a January 16,
2003 Pravda, RU story titled "Scary Secrets of the Third Reich's Base in Antarctica, 't where we learn a research expedition discovered a virus in Antarctica" that neither people nor animals had immunity to" and which could
"cause a catastrophe" to the planet. IS THIS FOR REAL??? The problem here is that there are too many disinformation stories being planted deliberately for mind control and scare tactics by Mr. Richard Hoagland, AS HE IS NOT A SCIENTIST, ONLY A PROFITEERING WRITER WHO
WANTS TO SELL MISINFORMATION TO THE HIGHEST BIDDER AT LARGE SO
THERE YOU ARE, MR. RICHARD TRICKSTER HOAGLAND!
THE MOST POSSIBLE SCENARIO COULD HAVE BEEN SOME DEVASTATION CAUSED AND CREATED BY ARRIVAL ON SOME METEORITES IN AREAS OF ANTARCTICA.

In the meantime, several nations — Russia, U.S., U.K.,
E.R.G., and Italy — have been working on plans and activities to drill into LAKE VOSTOK, beginning January, 2006. . . and other lakes since 2006... The sterile drilling and environmentally robotic exploration should be interesting... Unfortunately, no results have been released since 2006 pertaining to these exploration technologies... Scientists consider that these sealed ecosystems, which probably do contain life, may mirror conditions on Jupiter and Europa having different biological forms of life as we know it... ALSO, I
PREVIOUSLY MENTIONED MERCURY
AS SHOWING SIGNS OF "LIFE" - PLEASE CONSULT MY PREVIOUS BOOK ENTITLED
"GRAVITY, MATTER & SPACE TRAVEL", PUBLISHED
IN OCTOBER, 2006 FOR MORE DATA CONCERNING LIFE
ON GANYMEDE, WITH THEIR HUMANOID ENTITIES

LIVING THERE, WITH THEIR VERY ADVANCED "FLYING SAUCERS" HAVING VISITED .... THIS planet earth on many occasions in the past.

THE MARTIAN CITIES ARE LOCATED
UNDERGROUND, AS CAN BE SEEN IN MY

PREVIOUS BOOK IN YEAR 2006! THE MOVIE
"TOTAL RECALL" WITH ARNOLD
SCHWARZENEGGER BACK IN THE 90s

WAS AN INDICATION THAT MARTIANS DEFINITELY DO EXIST, THAT'S FOR SURE!

EXTRATERRESTRIALS, THE VATICAN AND EARTHLINGS, TRANSCRIBED BY Data!

THIS AUTHOR FROM DATA OF LATE ZECHARIA

SITCHIN MCRE NEW DATA DIALOGUE IN

BIMINI — BELLBRiA

SITCHIN + VATICAN THEOLOQAM m SCUSS VFOs,

FXTRATERRESTRTALS, ANGELS, _CRERTJOW OF MN in Year 2005.... In what must be a

historic first, a high official of the Vatican and the late Zecharia Sitchin discussed the

issue of extraterrestrials and the creation of man on earth, background, religion and

upbringing, both arrived at 3 common conclusions:

Extraterrestrials can and do exist on other planets in various dimensions, not only in the 'Visible spectrum' here on planet Earth!

They can be more advanced than us holistically and spiritually!

Materially, mankind could have been fashioned genetically from a preexisting sentient race of beings having lived and worked here. Furthermore, these 'reptilians' co—existed with the dinosaurs and plesiosaurs many millions of years ago in various areas on Earth...

These 'Anunnaki' reptilians or 'lizard' people had a
3rd eye on their foreheads and
2 'horns emanating/ from both sides of their heads. They were called the

'CHITAULT' in South Africa, residing in underground cave systems below the 'Drakensberge or Dragon mountains there and elsewhere on this planet.

Msgr. Balducci's Positions On UFOs. "There must be something in it. " The hundreds and thousands of eyewitness reports leave no room for denying that there is truth in them, even allowing for optical illusions, atmospheric phenomena, etc. As a Catholic theologian such witnessing cannot be dismissed. "Witnessing is one way of transmitting truth, and in the case of the Christian religion, we are talking about a divine revelation in which witnessing is crucial to the credibility of our faith. ON LIFE ON OTHER PLANETS -

As previously mentioned by this author in his book: "GRAVITY, MATTER & SPACE TRAVEL", the humanoid form is prevalent on many other planets especially in our own solar system and this is being emphasized here again. Life can exist on other planets in our Universe. . .The Bible does not rule out that possibility. On the basis of scripture and on the basis of our knowledge of God's omnipotence. God's wisdom being limitless, we must affirm that life on other planets is very possible. " Moreover, this is not only possible, but also credible and very probable. Cardinal Nicol o Cusano (1401—1464) wrote that there is not a single star in the sky about which we can rule out the existence of life, even if different from ours."

ON INTELLIGENT EXTRATERRESTRIALS: "When 1 talk about extraterrestrials, we must think of beings who are like us more probably, beings more advanced than us in spirit and consciousness, in that their nature is an association of a material part and a spiritual part, a body and a soul, although in different proportions than human beings on Earth. " Angels are beings who are purely spiritual, devoid of bodies, while we are made of spirit and matter but still at a low level. It is entirely credible that in the enormous distance between angels and humans, there could be found some middle stage, that is, beings with a body like ours but elevated more spiritually. If such intelligent beings really exist on other planets, only science will be able to prove; but in spite of what some people think, we would be in a position to reconcile their existence with the redemption that GOD has brought us.

THE REPTILIAN ANUNNAKI AND THE CREATION OF MAN AND WOMAN.

We appear to agree, Sitchin said, that more advanced extraterrestrials can exist, and use science to evidence their coming to Earth. Sitchin then quoted the Sumerian texts that say that the Anunnaki (Those who from heaven to Earth came") genetically improved an existing being to serve as their slaves on Earth to create the beings called 'Adam and Eve' in the Bible...

My conclusion regarding your presentation, Msgr. Balducci answered, is that more than anything else your whole approach is based on physical evidence, it concerns itself with matter, not with spirit. This is an important distinction, "because if this distinction is made, T can bring up the view of the great theologian, Professor Father Marakoff, who is still alive and is greatly respected by the Church. He formulates the hypothesis that when God created men and women and put the soul into them, perhaps what is meant is not that they were created from mud or lime, but from something pre—existing, even from a sentient being capable of feeling and perception. So the idea of taking a pre— man or hominid and creating someone who is aware of themselves is something that 'Christianity' is coming around to. The key is the distinction between the material body and the soul granted by GOD. 'I FROM REPTILIAN ANUNNAKI TO GOD

Sitchin dealt with the physical evidence but in his book 'The 12th

Planet', the very last sentence of the last paragraph raises the new question: If the extraterrestrials 'created' us, who created them on their planet?

The Anunnaki, Sitchin explained, were just emissaries — and that is what the Hebrew word "Malachim", translated 'angels, means. They thought that it was their decision to come here for selfish reasons and to fashion us because they needed workers but in truth they only carried out the Almighty God's wishes and plans.

If such extraterrestrials were so involved, Msgr. Balducci said, even by your own interpretation they had to do with man's physics, body and rationality: but COD alone had to do with the soul!

Sitchin told Msgr. Balducci, "It seems to me that we are ascending the same stairway to heaven, though from different steps, "Sitchin said. Finally, there will be numerous "earth changes" on a very large scale

— it translates to December, 2012, the exact date of the end of the "Fourth World" according to the Mayans. These "earth changes" actually started with the very powerful tsunamis of 2004, continuing on into other numerous, very powerful seaquakes, earthquakes and other very devastating natural disasters worldwide. I PREDICT THAT AT LEAST of THE EARTH'S POPULATION WILL BE TOTALLY DESTROYED BY GOD'S WRATH DUE TO THE DECEPTION, CORRUPTION
AND WARS BEING WAGED NOW THROUGH UPCOMING
YEARS, UNTIL THE END OF YEAR 2060. MAY GOD BE
WITH YOU, ALWAYS. CONCERNING "JESUS" AND THE
"BIBLICAL PATRIARCHS"
The Biblical Patriarchs were not simple shepherds they were called the "HYKSOS," the "Shepherd Pharaohs" of
Egypt.

ABRAHAM, JACOB, MOSES IN THE BIBLE, MAMAYBRA, JACOBA, TUTHMOSES IN EGYPTIAN HISTORY.

The Biblical Exodus is a historical event the Hyksos exodus from Egypt into Lebanon a very long time ago, about 000 BEFORE PRESENT NOW.

JESUS too was a Shepherd pharaoh prince, a 'Lamb of God'

Jesus was heir to the astrological theology of ancient
Egypt. The worship had moved over the millennia from Apis Bulls (Taurus) to sheep (Aries) to fish (Pisces). Therefore, Jesus began as a 'Lamb of God' but became the 'Fisher of Mankind'.

This is exactly what the Bible and other ancient texts in Sanskrit language tell us. Biblical history is not totally divorced from 'real' history, as it sometimes appears, in fact most of the Biblical heroes are to be found in the historical record.

The fairy—tale faded of Jesus living in France is NOT BASED UPON REAL HISTORICAL AND BIBLICAL FACTS. Furthermore, the 'SHROUD OF TURIN' is another fairy— tale story concocted by the Vatican and others of our 'Western religions'.

There is so much more data to follow in the data I'm transcribing, originally by another author friend, who passed away in 1989... THE NAME OF THIS LATE AUTHOR FRIEND WAS RICHARD J. BENSON, ALSO KNOWN FOR HIS
BOOK: "GOD'S 1 ANCIENT LANGUAGE SPOKEN IN
TONGUES", + more data +24 THE STORY OF SHAMAN
THE ANCIENT WISDOM OF INDIGENOUS PEOPLE

REVEALS A VERY STARTLING SIDE OF HUMAN HISTORY ON THIS PLANET EARTH:

This story is being transcribed by this author from 'Vusamazulu Credo Mutwa' a renowned Zulu elder and author of 'Song of the Stars: The Lore of a Zulu Shaman'

Anthropological orthodoxy insists that civilization began in Sumeria 6,000 years ago, and the modern metropolis is the pinnacle of culture and evolution on the planet. But, circa world war 2, humanity shattered the rails of our technological playpen, sporting new atomic bombs, originally developed by German scientists before world war 2. As I said in this book, space—faring ETS took notice, actually dating back about 100 years, and silver flying saucers suddenly filled the skies.

The 'UFO era' was born actually in 1891, 120 years ago, to be exact! An alternate view is emerging, however. According to indigenous people from the Americas to South Africa, they have guarded the hidden history of humanity all along, quietly maintaining contact with visiting and resident stellar relatives. Mobilized, now, by what they believe is the fulfillment of long— prophesied warnings, the elders of these indigenous people around the world have begun, they say, to break vows of silence and share their ancient secret stellar wisdom. AFRICAN EPIPHANY

Thanks to the work of Robert Temple (The Sirius Mystery), the start— ling knowledge of Sirius and its dwarf companion by Africa's Dogon tribe is widely known. The Dogon possess knowledge, such as the star system's orbital periods and the companion stars' invisibility, which cannot be confirmed by naked eye observation, and which modern astronomers have learned only recently.

Credo Mutwa claims and has chosen a path of 'openness', coming forward to share secret star lore of indigenous Black Africans. pray that this effort will unite thinking people around the world and diminish the severity of our prophecies, " Credo explained recently. A master storyteller, he has traveled to more than 20 countries sharing his vision and wisdom despite the great personal loss which his openness have cost him. Credo's son was brutally murdered by those who want him silenced.

Indigenous cosmology of stellar relationships is complex, he explains, often dwelling at the heart of sacred ceremony. Rich oral traditions, including protocol for contact and how to distinguish friendly off— earth visitors from those who are potentially harmful, have been handed down from one sangoma (shaman) to the next for thousands of years. Star lore is an essential component of sangoma training. According to Credo Mutwa, "In every language in Africa, the meaning of star is Bringer of knowledge, or Bringer of enlightenment." Credo has traveled the continent of Africa, sculpting haunting images of visitors from the stars, which were described to him by other African shamans. I' These beings have been coming and going to Africa for 40,000 years, t' he says. Some bear striking resemblance to beings reported by modern experiencers of the UFO phenomenon.

Cradled in South Africa's Krueger National Park lies a private game reserve called Timbavati. This emerald jewel of the African bush is almost mythical in reputation. White lions are born in here, it is said. "A long story is told about a chieftainess called Numbi, Credo re— counts. "Many generations ago, she and her people saw a burning white light like a star fall out of the sky right where Timbavati is today. The story is that it was not a star; it was a shining ball of metal, brighter than the Sun. When this ball came down to the ground, Queen Numbi, who was a sick old woman at that time, went towards the lights and was swallowed by the light. In that light, very faintly seen, were strange beings with very large heads. These beings received Numbi into the light, and for some hours she was inside. When she emerged and walked toward her people, she had become much younger than when she had gone into the light.

After that star fell, stayed on the ground for some days, and then rose back into the sky, strange things started happening there. Cattle with 2 heads were born repeatedly. Lions, leopards and even impalas with snow—white fur and green eyes were born, until to this very day. This story is one of the most amazing in Africa. Even to this day, white animals are still being born in Timbavati. Some years ago, a snow—white elephant with beautiful blue eyes and long tusks used to roam the area, until white hunters shot it.

"When a tribe of invaders appeared at Timbavati many years after Numbi experience, Credo said, "they brought sacred stones which had been taken from Zimbabwe, and planted these stones there in honor of that place. Timbavati, which is Zulu for 'the falling down of a star' is one of the holiest places in South Africa. But now its story is löst and has been overshadowed by a lot of nonsense.   The standing stones of Timbavati, brought from Zimbabwe to honor Numbi 's visitors, are reminiscent of megalithic sites around the world and give mute testimony to the antiquity of the place. Most of the stones now lie on the ground, overgrown by the grasses of the African bush, but the outline of a large circle is suggested.

The standing stones have a resonant quality when struck with a smaller stone, similar to the deep, bell— like resonance of certain Egyptian monoliths. Adjacent to Timbavati is an enigmatic place known as Many— eleti, which means 'Gateway to the Stars. ' A community of 30 shamans lives there because they believe Manyeleti binds heaven and earth. In his book, Credo Mutwa relates a prophetic
Vision of 4 great leaders emerging around the world: red, black, white and yellow.

The colors are the same as the Lakota Medicine
Wheel, mentioned below, and 4 races of humanity.

"These leaders will work to unite the planet," Credo says. "One of these, a female leader, will arise in America. She will be called the Red Savior, because of the fiery color of her hair. " Native Americans, such as the Lakota Sioux, have an expression, "Mita— kuye Oyasin, " which means "All our relations." Four— legged ones, winged ones, crawling ones, plant and stone nations are greeted as relatives. The Lakota Medicine Wheel is composed of red, black, white and yellow, representing 4 races of humanity. Within Native American cosmogony, it is natural to include and respect the Star Nations among extended family members. Standing Elk, Dakota Elder and Sun Dance chief, recently presented an open letter to the Elders of Turtle Island. "My heart told me to speak of the secret knowledge of Native Americans concerning the Star Nations, since the time of our prophecies is at hand." Believing the knowledge belongs to the world, Standing Elk has created Star Knowledge

Gatherings, a forum to share this information. Sharing such secrets is controversial and unpopular with some native peoples. Standing Elk, like his African counterpart, has received numerous threats.

At Standing Elk's gatherings, Native Elders share the conference podium with prestigious researchers in the UFO field. "Alien" contactees speak openly of their experiences. Indigenous Elders perform ceremony and give candid testimony of their knowledge and relationship to the Star Nations.

"Humanity was seeded from the stars, and we have a profound genetic kinship with humanity's stellar brethren," Rod Skenedore, a native elder, recently told an audience of hundreds in Las Cruces, New Mexico.

At the same Star Knowledge Gathering, Quiche Mayan Elder, Grandmother Windrider, talked of her visions and her challenging personal path. "1 challenge conscious people to stand together in peace and tolerance to make a difference at this time on Earth", Windrider said.

MAYAN CALENDAR CONNECTION

From Central America, Guatemalan Mayan Elder, Don Alejandro, speaks of vast and repetitive cycles of time. "The Mayan calendar never has to be altered, because it

is based on the stars," he points out. Our unwieldy Western counterpart, on the other hand, has been changed many times and is still not accurate. " This knowledge was bequeathed to the Mayas by the Abuelos, the grandfathers, who came from the stars," Don Ale— j andro said. He enigmatically links the origin and destiny of the Mayas with the Pleiades, who he says, were called May. According to the work of archeoastronomer Anthony Aveni, certain Mayan sites appear to be aligned with zenith risings of the Pleiades. One of the named stars in the constellation of the Pleiades, the 7 Sisters, is Maia. December 21, 2012, will be the end of the current Mayan long count,

According to scholars, this Great Cycle began August 11 r 3114 BC. in Mayan terms this time period equals 13 baktuns. In the Gregorian lexicon, this equals 5125 year. 5 great cycles, or suns, equals 25,625 years, which is amazingly close to what modern astronomers count as 1 cycle of the precession of the equinoxes, 25,920 years. Mayan day— keepers say we are living in the time of the 5th sun, approaching the end of a major cycle. Mayan prophecies also point to this time as filled with major earth changes and transformations.

On December 21, 2012, winter solstice in the Northern Hemisphere, a significant astronomical event will occur. From the vantage point of Earth, looking through the constellation of Sagittarius toward the center of our Milky Way galaxy, our Sun will align the Galactic Center. This close conjunction occurs only during a specific epoch each 26,000 years. In his book 'Maya Cosmogenesis', John M. Jenkins likens the 4 quarters of the precessional cycle to seasons of a grand year. He believes this juncture is the completion of 1 full cycle, described and monitored by the sophisticated Mayan calendar.

PROPHETS AND SEERS

Ancient prophets and modern seers seem to speak with one voice and see with a single eye. The Bible, Nostradamus, Edgar Cayce, Hindu texts and Hopi prophecies predicted this time on earth would be marked by miracles and cataclysms, the great storms before a long—prophesied golden age. Credo Mutwa said, "Anyone who investigates will come upon this amazing fact. In South America, in Brazil, in Peru and in Bolivia, different Native American tribes are expecting worldwide cataclysms in this coming century. They have been expecting these cataclysms for centuries, and they knew it would occur very early in this coming 21st century."
If numerous tradition occurs that we 're poised at the culmination of their prophecies, how did they know the specific time frame thousands of years ago? If the triggering of global cataclysms is an external object like a comet, with a cyclical orbital period, its return would be predictable and monitoring the movements in the sky would be vital. The last such juncture seems to have occurred roughly 13,000 years ago, halfway through this large cycle and resulting, it would seem, in the last ice age.

Pieces of this puzzle continue to move into position. Investigators increasingly believe that our history, and that of our progenitors, has been recorded in legend, myth and star lore. William Sullivan's ground— breaking work with the Incas, (Secret of the Incas), decoded myths as astronomical metaphor. Robert Bauval and Graham Hancock have shown that megalithic monuments contain critical stellar alignments and, when understood, convey messages and timing significance.

Physical evidence of earlier life on Mars, now with the 7 underground cities of the Martians having been shown previously by this author!

Other forgotten civilizations, including Atlantis and Lemur i a, among others worldwide, certainly preceded us their life cycles and attainments terminated by cataclysmic events which also destroyed the evidence. Humanity's true origins and history may have been bequeathed to us through the metaphorical oral traditions of Earth's indigenous peoples and a world—wide web of stones, including the world—wide pyramid—grids extending also into

China with its 'GREAT WHITE PYRAMID' this grid extending to planet MARS, with its 7 pyramid cities!

Sherlock Holmes asserted that the best place to hide something is in plain sight. It appears that once we have decoded these Rosetta Stones of the stars, our place in the interplanetary scheme may become more clear. When we remember the codes and decipher all of the messages, the truths may very well be obvious. THE MYSTERIES OF MARS

Imagine that you returned home to find the house next door a burned— out, lifeless wreckage, uninhabitable and abandoned by the former occupants who were never to be seen again. If you're like most of us, learning what actually happened next door would be vitally important.

Now consider that scenario played out on a cosmic scale.

Mars once had an ocean larger than the Pacific. Mars still has an atmosphere on its surface, along with very extensive canals of water all over and vegetation, including numerous biologic forms of life, as was previously mentioned in my book "GRAVITY, MATTER & SPACE TRAVEL" in 2006. What happened on Mars, our neighbor and could what happened on Mars happen to us here???

Like it or not, humans are performing wild experiments on the atmosphere, oceans and climatic systems of our own planet, and since we have so little knowledge of the outcomes on planetary scale, these lessons that Mars teaches are vital for us. Mars is Exhibit A. However, when images from the 1976 Viking Mission revealed features on the surface of Mars that look remarkably like ruined architecture in the Cydonia region, an amazing independent scientific inquiry began that continues to this day. The question now became: Could Mars have sustained conditions for life to exist? Jet Propulsion Laboratory's new paradigm was a view that Mars' history was far more Earth—like: wet, warm and alive.

In 1986, Dr. John E. Brandenburg, a long—time
Mars researcher, came out with the book 'DEAD MARS, DYING EARTH', became the first scientist to stand before a scientific conference and announce the hypothesis that a paleo—ocean once existed on Mars and still exists there today!

Recent visual evidence from the Mars Global Surveyor of past flowing water and of an ocean shoreline supports his claim. Perhaps the most telling evidence of the paleo—ocean has been from the recent images of Mars produced by the Mars Orbital Laser Altimeter (MOLA), which shows a huge topographical depression in the Northern Martian Hemisphere where the ocean, covering about 1/3 of the surface of Mars, was then located.

The presence of this large body of water on the surface of Mars tells us a great deal about the planet's history. For example, the presence of lots of liquid water reveals the temperatures on Mars, currently ranging from —137 0 to +27 0 F, were at one time above freezing. There is new evidence also that 'sand— whales' have been seen all over this huge body of water• Please refer to the photos in my volume "GRAVITY, MATTER + SPACE TRAVEL", SHOWING THESE BIOLOGICAL LIFEFORMS IN THERE!

THIS IS A VERY REMARKABLE SCIENTIFIC FACT NOW. The presence of flowing water and warmer temperatures above freezing now indicates an atmospheric greenhouse and atmospheric pressures.   Atmosphere on Mars today is about 10% of Earth's. This atmosphere was probably primarily carbon dioxide, which is a very effective gas for retaining solar energy and warming a planet. However, the red color of the Martian soil tells us that there was abundant oxygen on Mars as well. The reddish color is caused by oxidized iron in the Martian terrain, like rust, and is Very similar to landscape seen in the desert in southwest North America and the cracked deserts left by the destruct ion of Amazon rain forests.

But was there life on Mars? Mars has actually sent a few rocks to Earth for us to examine.

Mars sent Earth a love note 16,000,000 years ago. Inscribed on a big stone, toward the starry ether thrown, it took millions of years to be captured by gravity to take a flaming plunge through the currents of our planet's atmosphere. Fallen and alone, it lay on ancient ice unread for 13,000 years, but when its code was broken, and its legend read, the message that was etched in stone told humans we 're not alone.

FROM BOOK DEAD MARS, DYING EARTH

It's difficult to imagine how a rock from the surface of Mars could land on our own planet. However, although we'd never brought back a sample from Mars, we had run tests on the surface during the Viking Mission that allow us to read the unique signature of oxygen isotopes in meteorites from Mars. The finding that the meteorite ALH84001 — which was discovered lying on the ice in Antarctica we were able to deter— mine that it was of Martian origin. Recently, the finding that it contained Mars bacteria was strongly bolstered when a form of magnetite only produced by living processes was found in ALH84001.

Did you know that Richard Hoover of the Space Science Laboratory at the George C. Marshall Space Flight center opened a space rock and found tiny mushroom— like forms inside which immediately turned to dust before his eyes? Fortunately, there are pictures that remain.

Evidence of life aboard meteorites isn't all Martian life. We 're discovering that there is a lot of life out there in the Universe! We seem to find evidence for it everywhere we look.

PANSPERMIA deals with notions about life from outside Earth's biosphere have never been popular in 'scientific circles' but LIFE DOES EXIST on some of our other planets and moons in our solar system and else— where, as stated by Eh is scientist! Nevertheless, y our understanding that life can survive under very extreme conditions has changed a great deal in the last 20 years. We now know that here on Earth it is possible for life to thrive deep within hot—water vents at the bottom of the ocean, deep under Antarctic ice, in volcanic lava and even in— side nuclear reactors. Does it remain inconceivable that life could survive in very extreme conditions in outer space?

This is a particularly interesting question when you consider how Mars + biological life on it died. It appears that water stopped flowing on Mars about 500 million years ago. This coincide with the very large asteroid strike, which created the Lyot impact crater on Mars, The Lyot impact was similar in size to the Chicxulub impact on

Earth 65 million years ago in the Yucatan region of Mexico. That strike threw so much debris into our atmosphere that it altered Earth's climate for long enough to cause a mass extinction. That's when the dinosaurs were wiped out on Earth. The 'plesiosaurs' survived in the oceans and still thrive today; some have been sighted on numerous eye—witness accounts, especially even inside some of the big rivers within Australia... However, eventually Earth's biosphere recovered, and the mass extinction created new niches for furry little mammals, including our hominid ancestors, to flourish, as previously mentioned by this author in his other books. However, Mars, being smaller and further away from the Sun, was not able to recover. Eventually, it lost most of its water on the surface ocean, leaving it a frozen wasteland, ultimately to achieve a second stable state, the one we know today.

Asteroid strikes continue to be a problem for planets and not just on Mars, which is particularly vulnerable because of its location near the asteroid belt. The asteroid belt is the leftovers of planet MARDUK or MALDEK, which literally imploded through a bombardment of several meteors, resulting in the various asteroids. The largest asteroid is Ceres, about 300 miles in diameter. As a human on Earth, your chances of being killed by a meteor, comet or asteroid impact is about the same as that of dying in an airplane disaster - 1 in 20,000. This isn't because the strikes are that frequent, but rather, because they are so devastating on those rare occasions when they do make contact. And as rare as they may be, our generation here on Earth was witness to a major strike on planet Jupiter when, in July of 1994, the Shoe— maker— Levy asteroid plowed into that planet one huge hunk after another, like a slow—motion train wreck. Now that we have become citizens of the Cosmos, aware of this threat to our home world, we need to expand our ability to deal with such a dangerous reality. Fending off a major strike is one of the few things that humans may be able to do to help all of the species of Earth. Project HAARP in Alaska can be used to implode any errant comets or meteors venturing into the vicinity of Earth. Another HAARP system is set up and operating in Russia. Not that humans are contributing much to the protection of Earth's biosphere at present. In fact, it may not require an asteroid impact to generate a catastrophe here on Earth. We t re clearly enacting a similar kind of planet— wide climate change by altering the chemistry of our own atmosphere. The problem on Earth is unprecedented levels of atmospheric carbon dioxide, the same greenhouse gas that kept Mars warm enough for liquid water which is still there in great quantities!

Many people aren't aware of the fact that carbon dioxide, which is a major by— product of the burning of fossil fuels such as oil, natural gas and coal, was considered a deadly gas until very recently in our history. Miners in the early 1900s and now called it blackdamp or carbonic acid gas, and respected it as a substance that regularly killed miners. Brewers in the past knew that carbon dioxide could be lethal and avoided placing their heads lower than their vats since the toxic gas would tend to accumulate toward the floor. However, in the 1940s, when human respiration was more fully understood, we determined that carbon dioxide was one of the by—products of our own bodies, our attitudes began to change. The fact that i E is also the substance that makes soda pop fizzy also hides the darker, more destructive aspect of its nature.

In fact, the darker side of carbon dioxide was once again demonstrated recently when asthma researchers showed that people in the midst of asthma attacks are exhaling highly acidic water vapor, in other words carbonic acid, the gasified combination of water vapor and carbon dioxide. Fortunately, they also think this implies an effective treatment for what has become epidemic of asthma. STAY AWAY FROM IT, PEOPLE! Carbon dioxide is the major component of the atmosphere of Mars. If you think of it as the substance of the atmospheres of played—out planets, you begin to see what the problem is with putting billions of tons of the stuff into our oceans and air every year!

We have already made fundamental choices as a species. For example, we live in a golden age of transportation, telecommunications and computer wizardry' Yet every time we bring home a new box of computer gear and plug it into our surge protector, we 're connecting to a grid of wires that leads back to a power station. We seldom consider that the technology behind the scenes is over 100 years old.

For the most part we still produce electricity using steam—driven turbines. About 75% of the time, the fuel being burned to turn those turbines is coal. So, your brand new mega— megahertz machine is essentially coal powered. Our space— age technology is connected to an antiquated method of energy production. Even if your electricity is produced by a fission nuclear reactor, that, too, is a technology that is over 50 years old and we 've never managed to get the bugs out, especially with all of the nuclear toxic wastes polluting many areas worldwide. THIS IS A TOTALLY
INSANE SCENARIO WORLDWIDE, ESPECIALLY WITH THE
LATEST NUCLEAR EMERGENCIES IN JAPAN AND
NUMEROUS OTHER EARTH CHANGES ADVANCING UPON
THIS PLANET AT ANYTIME AS SOON AS POSSIBLE WITH SO MUCH DESTRUCTION! Not much better can be said of the combustion engines and hybrids that drive our cars on a daily basis...
So we have to ask ourselves: Are we so married to these shamefully out— dated methods of energy production that we 're willing to risk becoming another lifeless rock with an out—of—control greenhouse? We have advanced in nearly every other area of technology; why don't we take on the task of updating our energy production methods? SOLAR TECHNOLOGY IS A VERY GOOD ALTERNATIVE TO DEAL WITH TN MANY COUNTRIES WORLDWIDE...

Wind—turbines are extremely inefficient and very costly to manufacture, with their overall efficiency rating at only 12% and they destroy birds and other animals, along with polluting environments on a global scale. OUR SMALL TEAM OF   ENGINEERS IS PRESENTLY INVOLVED IN PROJECT "MAGNETIC GENERATORS + MAGNETIC MOTORS ON A LARGE SCALE" TO IMPLEMENT RADICAL IMPROVEMENTS PERTAINING TO NEW ENERGY TECHNOLOGIES WORLDWIDE... WHY AREN'T WE DEVELOPING FUSION ENERGY, A MUCH SAFER + MORE EFFECTIVE FORM OF ENERGY? THIS SCIENTIST HAS THE "SAVOIR-FAIRE" TO MANUFACTURE A PILOT FUSION PLANT BY YEAR 2012, USING A PLASMA-FUSION CHAMBER TO MANUFACTURE SAFELY BILLIONS OF KWS PER HOUR WORLDWIDE!

We need to look very carefully at the choices we are making by doing nothing or little, by conducting business as usual because without change, Earth may soon become another SINKING TITANIC! THE CHOICES ARE YOURS, LADIES + GENTLEMEN, AS WE FACE MANY CHANGES AT THE ENERGY SECTORS WORLDWIDE IN
VIEW OF THESE CATASTROPHES ABOUT TO HAPPEN IN THE VERY NEAR FUTURE! THANK YOU

FOR YOUR ATTENTION CONCERNING ALL OF THESE IMPORTANT SCIENTIFIC MATTERS DISCUSSED HERE TN THIS BOOK!

BACK TO THE FUTURE, AS THE FUTURE TS HERE NOW. • •
Hans J. Petermann,

SCIENTIST + LINGUIST + ESOTERIC + EXOTERIC
RESEARCHER FROM AUSTRIA + CALIFORNIA, UNITED STATES OF AMERICA... MAY GOD BLESS YOU ALWAYS. . .

THE COMING SOLAR SHIFT + FUTURE NATURAL
DISASTERS

If the Venus Transit, Phoenix Cycle and 2012 calendar are predictive that portend a radical shift in solar output, accompanied by earth changes, then we should find evidence of this in the recent geological and climate records. Some scientists blame carbon dioxide emissions and the resultant 'greenhouse effect' for recent global warming. How— ever, there is no doubt that the sun is the main forcing mechanism responsible for the 13,000 year interglacial we're still in. When the sunspot cycle shut down in 1640, the 'little ice age' soon followed. As sunspots have increased over the last 3 centuries the Earth has heated up, glaciers have melted and the sea level has risen.

The Sun is the source of life, light and energy on Earth, and this ought to make us pause for a moment upon the wisdom of the ancient solar religions. When solar output changes, the earth's magnetic field and climate also change. Sunspots are concentrations of magnetic flow through the sun's surface. They typically occur in clusters with a cyclical fluctuation of 11 years. As sunspots increase, so does solar activity, including flares' the most violent events in the solar system. Researchers into a wide variety of phenomena including the stock market, social unrest, wars and natural disasters have uncovered links between the sunspot cycle and the fluctuations of these known phenomena.

Using the late 17th century as the zero point of this oscillation, we see that the actual peak of the 300—year cycle of increasing solar output occurred in 1958— 1959. The number of sunspots was about 220; the usual peak is around 100—150. Now, what's interesting is that during the first of the 20th century the earth's seismic and volcanic activity were comparatively quiet. Then in 1960, a magnitude 9.2 super—quake rocked Chile in South America. killing around 10, 000 people. From that time on, the level of seismic and volcanic activity increased steadily to the point that the 1990s can accurately be called the decade of disasters.
The surge in major earthquakes and volcanic eruptions radically departed from earlier decades.

According to the chief scientist for the world's largest reinsurance company Zurich Re, "since 1960 natural disasters are a growth industry.

The company is in the business of insuring against natural disasters so they carefully monitor and report on the 'real' trends. T make this point because the government and many scientists claim that there has not been an actual increase in the total number of earthquakes. That may be true but it is not the total number we are concerned with but the number of potentially dangerous high magnitude earthquakes (natural dåsasters).

According to USGS figures, the frequency of occurrence of earthquakes analyzed from data accumulated since 1900 shows that there were 120 strong (6—6.9) tremors, 18 major (7—7.9); and 1 great quake (8—9.1) on an average annual basis. A chart for the number of significant earthquakes worldwide from 1990— 2000 is more revealing of recent trends. In 1994 there were 161 strong, 13 major and 2 great earth— quakes. In 1995 there were 185 strong, 22 major and 3 great earth— quakes. In 2000 we find 153 strong, 16 major, and 4 great earthquakes. Those numbers reveal a startling, statistically significant jump in the frequency of dangerous tremors!

We find virtually the same kind of data when we examine the recent upsurge of volcanic eruptions and severe storms. Hurricanes, tornadoes and massive floods have increased in frequency and violence since the 1980s, facts that the mass media reports on every year. As the sun— spot cycle hit its second peak an amazing 20 volcanic eruptions occurred around the globe on 1 day, August 26, 2002. The 2012 end of the 5th Sun — prophecy includes an increase in natural disasters and the record strongly supports this ancient prediction.

In the past several decades we have learned that our 'electromagnetic' civilization is very vulnerable to intensified solar activity. Where we once imagined that there was a comfortable void between the earth and the sun, we now know that this space is actually a maelstrom of solar wind particles and the 'Van Allan' radiation belt, ultraviolet and x—rays streaming toward our planet. The solar wind is so potent that it bends the earth's protective, magnetic envelope causing it to quiver. When the sunspot cycle is at its peak solar flares erupt and routinely knock out radio & telephone transmissions, satellites and they can even disrupt the power grid.

While the governmental and scientific institutions downplay the reality of increasing solar activity and the related rise in natural disasters, scientists quietly acknowledge the threats. The ability to predict earthquakes and volcanic eruptions has quickly risen to the level of a science in the past 15 years, At the same time new theories based upon, heretofore, taboo ideas that claimed the earth occasionally changed, due to massive cataclysms suddenly came into vogue. But are these theories really new? If the oral traditions and written histories of ancient civilizations, are carefully examined one can only conclude that the ancient solar priests were well aware of the fact that the earth goes through cyclical transformations. My request is for an objective examination of the accumulated evidence, which appear to validate the ancient science.

I do not think that the Maya picked 2012 by accident. It not only coincides with the next sunspot cycle peak but with the 2nd leg of the Venus transit. The 300— year solar upswing from the

17th century (zero point is self-evident. It appears that the oscillation peaked in 1960. Solar activity has remained high and yet the sunspot peaks have been declining since then. The key unpredictable variable at this point is the Venus transit. The scenario that appears to be shaping up is completely contrary to the extended 'global warming' predictions.

If the Venus transits trigger a significant decrease in solar activity, we can expect a global 'cool down' or even a mini—ice age. This can happen much more rapidly than people assume as was demonstrated in the 1600s. While the global warming hype dominates the press some maverick' scientists disagree with the scenario. In an article titled Quieter, Longer Solar Cycle #23 Could Signal Significant Climate Shift' meteorologist Joe d' Al eo wrote. "If indeed the lower maximum and longer cycle is indicative of a NEW trend down in solar activity, global temperatures could soon react in the same direction down. Historically quieter periods have been associated with colder than normal temperatures.

The ancients — prior to the Greek and Roman periods
— had a paradoxical view of the planet Venus. She

was not simply the goddess of love and beauty, Venus was also the goddess of war and disease. The Maya feared the planet and so did the Sumerians; as Sekhemet she visited plagues and suffering upon the Egyptians who dared to ridicule the aging Rae If the priests were referring to the Venus transit as representing 'Ra's glance' then we should also pay attention to the other parts of the predictions. Those involve impacts by cosmic objects. Is it merely a coincidence that the Shoemaker—Levy comet smashed into Jupiter in 1994, to the astonishment of astronomers? In recent years several asteroids have passed close to the earth, in astronomical terms. These events have served as 'wake up' calls for many scientists who are now carefully monitoring the heavens in search of any 'incoming cosmic rocks'. The ancient science clearly predicted that we would enter this 'at risk' time frame near the end of the solar cycle. The Maya said the 5th Sun — the Sun of Motion would culminate in earth—quakes and impacts by celestial objects.

The Egyptian and Mayan priests knew that the sun went through oscillations, died and was reborn from its own ashes, like the Phoenix. They positioned the Morningstar and Venus transit in association with the sun and the solar cycle. This pivotal movement in human history is fraught with danger but it also presents a rare opportunity. As we enter the portal 2011—2017 of physical transformation we have the chance to shift our own focus from the blinding materialism of our self—centered civilization and emerge spiritually cleansed into the light of the new Sun!

As mentioned previously by this author, nuclear fusion can be created by construction of a plasma—fusion, hydrogen—helium chamber, with temperatures ranging at 100 million degrees Fahrenheit, at relatively safe environmental benefits for many countries worldwide. The cost for construction of a pilot plasma—fusion chamber would be around $ 40 million, output 40 million kilowatts/hour. It can be built safely within 1 year...

This is a product of the criminals working at Los   Alamos National Laboratory in New Mexico, U.S.A. Weapons laboratories in many other countries produce very similar products.

THE POTENTIAL OF THESE VAPORIZING PRODUCTS TS MEGADEATH, ECOCIDE + CLIOCIDE (THE END OF HISTORY). Weapons workers are "good" people, "only following orders" as all soldiers do throughout these insane world governments! YOU ARE RESPONSIBLE. Every person on this planet should do something to take away from world leaders this life threatening and despairable potential for destruction! ONE BOMB IS TOO MANY!   KIND PERSON,

Should this fate be yours or mine or anyone's? TO BE OR NOT TO BE, THAT IS THE QUESTION. Into the hands of a few less than noble God—men has been thrust the power to DESTROY US ALL. Our earth is being held hostage by all of these government criminals worldwide. POLITICAL INTOLERANCE! Ay, there's the rub. To see and stand the path to one's own crematoria and fain unconcern. What say you to the charge that you do naught to allay the very real and present dangers?

No traveler returns this path except as smoke. And, that smoke is this cause for lethal refrigeration of the rest. WOE IS ME! 'Tis an aware and alert citizen who sends this missive to you all! Let not this appeal be to a dull mind.

IT CAN HAPPEN TO YOU BY ACCIDENT OR INSANE INTENT.

This was published originally by Edward B. Grothus, Box 795, LOS Alamos, New Mexico 87544 USA/ for the 66th anniversary of the bombing of Hiroshima, Japan on August 6, 1945.

Upon very close examination of enclosed atomic bomb' explosion at Bikini atolls, you can actually see 3 'UFOs' viewing this scenario and exiting into space...

ASTEROID HIT!
JUPITER ONCE AGAIN SCARRED

Once again, as if to emphasize its cautionary message to nearby planets, Jupiter has been struck by debris from space, and as with the notorious Shoemaker—Levy event in 1994, astronomers on Earth were watching closely. This time camera on the newly refurbished Hubble space telescope were deployed and were rewarded with a spectacular view of the immense scar, left by an event which can only be described as a cosmic catastrophe in year 2009! Unlike the meteor crater event that struck Arizona, however, an impact with an object such as fell to Jupiter could have spelled doom for our little world, once again, the world has been shown that history in this solar system does not always unfold slowly and gradually.

Early in July, 2009 the world's largest telescopes were trained on the largest planet in the solar system. Not to miss the potentially new science in the unfolding celestial drama 580 million kilometers away, Matt Mountain, director of the Space Telescope Science Institute in Baltimore, Maryland, allocated discretionary time to a team of astronomers led by Heidi Hammel of the Space Science Institute in Boulder, Colorado.   The Hubble picture, taken on 23 July 2009, is the sharpest visible light picture taken of the feature and is Hubble's first science observation following its repair and upgrade in May. Observations were taken with Hubble's new camera, the Wide Field Camera 3 (WFC3).   Discovered by Australian amateur astronomer Tony Wesley on July 19, the spot was created when a small object plunged into Jupiter's atmosphere and

disintegrated. The only other time in history such a feature has been seen on Jupiter was in 1994.

"This is strikingly similar to the comet Shoemaker—Levy that impacted Jupiter in July, 1994," said team member Keith Noll of the Space Science Telescope Institute.

Planet Jupiter has an atmosphere of 67%, the rest consists mostly of rare gases, contrary to what nitwit scientists at 'N.A. S.A. ' have been spreading with their misinformation & disinformation campaigns! REMEMBER = N.A. S.A. - 'NEVER A STRAIGHT ANSWER', due to their schemes of manipulations, going back many years. if you'd like to read a very good book by Bill Kaysing, 'WE NEVER WENT TO THE MOON', please try to obtain a copy of it! There are actually 2 space bandits/ organizations!

Shoemaker's discovery caused a sensation in the geological world, as it was the first definitive proof of an extraterrestrial impact on the earth's surface when the meteor crater was formed near Flagstaff, Arizona in 1905. Since then, numerous impact craters have been identified around the world. In fact, the second suspected meteor crater was recognized in 1926, in Odessa, Texas.

Today, there are 175 known meteor scars on Earth, but Meteor Crater is one of the most famous and easiest to visit. The dry climate that reigns over this region has also caused little erosion. Travel guides to Arizona highlight Meteor Crater, but don't list it as a must. Instead, they argue that the Grand Canyon and Monument Valley are far more important to your visual sense than Meteor Crater. Though they are right, they also miss the point:

Meteor Crater is literally on par with the biblical warnings that GOD could strike whomever down Sodom and Gomorrah for example in an instant. Meteor Crater proves you don't need God's wrath to create instant deaths worldwide! Once on top of the rim of meteor crater, one can see a very impressive 175—meters—deep crater, more than one kilometer across in the desert! THE LATEST DATA CONCERNING THE FAR SIDE OF THE MOON
IS THAT IT HAS AN ATMOSPHERE OF 70% COMPARED TO OUR PLANET. The humanoid inhabitants of the moon mostly live in underground facilities and dome—sized buildings on its surface, well camouflaged. There are lakes, rivers and various kinds of vegetation, including trees in existence there•. THE CONSTANT
DISINFORMATION BY "N.A. S.A. MUST STOP! BILLIONS OF DOLLARS HAVE BEEN SQUANDERED BY THIS CRIMINAL ORGANIZATION THAT ONLY WANTS TO
MILITARIZE THEIR SPACE VENTURES AS IT PERTAINS TO
THE MOON AND OTHER MARS EXPLOITS. • • YOU REALLY
NEED TO STUDY MY PAST EPIC BOOK ENTITLED 'GRAVITY, MATTER & SPACE TRAVEL' for further enlightenment concerning the 7 underground cities on Mars and their vast underground transportation networks and cities all the way down to the south polar area, near it you can view the 'woman in granite' smiling at you, as you approach her from space with your 'flying saucer', coming in for a landing at "CYDONIA" pyramid cities complex locations near the other Mars face! There is also a photo of 'spider—man' looking at you from 'shadowy spider— rocks', along with photos of 'sand— whales' in the Mars waters! Finally, when will our next 'Mars mission' take off from this backward planet Earth? LET'S GO BACK TO THE FUTURE, AS THE FUTURE IS HERE NOW! YOUR BODY'S MANY CRIES FOR WATER by Feyedoon Batmanghelidj M.D.

This is a very good transcribed chapter written by the scientist known as 'Trevor James Constable. He is primarily known for his 'cloudbusting technological ventures' based upon the late scientific genius Wilhelm Reich's works. Reich was known for his 'orgone' energies developments.

RARE indeed are those books destined to become all— time classics. Even rarer are books destined to accomplish a paradigm shift in any major area of modern knowledge. Of still greater rarity are books destined? To benefit significantly the health of countless millions of humans, at no cost to them.

Dr. F. Batmanghelidj is a London—educated Iranian medical doctor who has made revolutionary discoveries about the water metabolism of the human body. His astounding basic breakthrough was made while he was confined to a Teheran prison, this being a unique circumstance.

Dr. Batmanghelidj is of aristocratic lineage in his native Iran, and when the Shah was overthrown, the doctor was arrested and jailed with more than 3,000 other well— born victims of the Khomeini revolution. While these unfortunates were awaiting execution, the Doctor was assigned as their medical officer — pending his own appearance before a firing squad. He had no medical resources other than water, in an environment pervaded by stress and terror. Indeed, the doctor found himself incarcerated in a gigantic stress laboratory.

This became the milieu in which fundamental discoveries were made regarding the medicinal and functional value of water. These are discoveries that have eluded giant medical trusts, vast hospital complexes, battalions of medical professors, universities that boast about their sophisticated research facilities, and all the vaunted resources of the pharmaceutical industry. None of them, or indeed, all of them combined, were capable: of penetrating to the bedrock of human health: adequate daily water intake.

Without realizing it initially, the doctor was working with clinical controls in place.
Prison discipline enabled him to follow up his patients, who had no possibility of evasion. Forced to use water medicinally, and water alone, the doctor was astonished in following up his patients, to find that water was effecting full cures of diverse, normally ineradicable diseases. These cures occurred in a complete fashion not seen in response to medication, which "treats" or "controls" such recalcitrant and diverse diseases as asthma, high blood pressure and ulcers. Official medicine has only palliatives for these conditions, not cures.

If such diseases do not bother you personally, just ask anyone so afflicted how frustrating it is to be perpetually treated but never cured, by doctors who are manifestly flummoxed by such diseases. This bafflement and medical inability to effect cures is almost universal in orthodox medicine & the legalized drug mafia, and largely accounts for the stupendous and ever—mounting costs of medical care. PEOPLE GET TREATED, NOT CURED! They aren't cured because the basic cause of their diseases is not known to classical medicine. AMERICA HARBORS AND HUSBANDS A VAST ARMY OF PEOPLE RECEIVING "MEDICAL TREATMENT".

Dr. Batmanghelidj was blessed by a first class medical training at St. Mary's Hospital Medical School of London University, one of the most respected medical schools in the western world. He was one of the last students of the eminent discoverer of penicillin, Sir Alexander Fleming. Thus steeped in classical orthodoxy, the doctor was totally embarrassed to find what water was doing in a dependable way, what medication had never been able to do. In his classical medical training, like orthodox doctor's worldwide, he was taught that it was material in the body the solutes were only an incidental status assigned to the solvent aspects of the human body — to water.

11 refutable clinical experience in the controlled prison environment forced the doctor to conclude that conventional medicine of which he himself was a product - was hobbled and handicapped by. The false paradigm undergirded, in place the illusion that water was not significant in human metabolism compared with the solutes, with which medical "science" has been vainly wrestling, generation after generation. The truth of basic human functioning, as the doctor's research confirmed progressively, is that the solvent (water) and not the solute, plays the cardinal role in human health.

Orthodox medicine has been very ineffective in dealing with a wide spectrum of diseases, whose etiology is classically listed as "unknown, ' because it has this fundamental polarity of solute and solvent back— to— front. The ramified and very fundamental role of water has simply not been comprehended. After 3 years doing basic research in the Teheran prison, Dr. Batmanghelidj was released, and came to America. Here he continued his pathfinding research on the water metabolism of the human body for 10 years or more. He worked with established institutions like the University of Pennsylvania, lectured widely to physicians, and wrote learned papers for professional publications such as the Journal of Gastroentology. The fallout from the doctor's exertions include a large number of completely cured persons who had been chronically afflicted with 1' incurable" disease. Medical doctors were among these people.

Slowly but surely, numerous intelligent medical doctors and researchers who studied his work, overcame the same embarrassment that had consumed Dr. Batmanghelidj, when he first began comprehending the cardinal role of water in human health and functioning. Like Dr. Batmanghelådj these enlightened physicians came to realize that medicine was con— fronting a fundamental paradigm shift. Some of these board— certified physicians are recognizing this central truth as time moves on.

Nothing so radical, in the root—meaning of that word, had ever happened to medical knowledge in the history of the world, chronic dehydration was being identified as the fundamental matrix from which most human diseases emerge. America's deplorable and desperate health care crisis is born of dehydration, midwifed by medical inability to deal e ct iwith diseases whose etiology in dehydration is not understood. Almost every disease arising out of dehydration is attacked with medications, or combinations of medications. Sickness therefore proliferates, while the sick cry out in many different ways for water cries to which doctors are deaf!

Countless ingrained cultural and social customs of modern America, such as not drinking sufficient water, and guzzling phenomenal quantities of diuretic soft drinks and diuretic coffee, reinforce chronic dehydration as the #1 health menace in America. We have to emphasize that drinking coffee, soft drinks and juices does not counter dehydration or meet the body's water needs. The only way out is by drinking water with a ph of 6.8—7.0; p, h. positive hydrogen. Also, $H4O2 = H4$ has 4 hydrogen atoms + 2 oxygen atoms in this great water!

Elusive and seemingly unrelated conditions like stress and depression, high blood pressure, dyspeptic pain, high blood cholesterol, excess body weight, arthritis, chronic fatigue, asthma and allergies, insulin independent diabetes, rheumatoid arthritis, back problems and many other complaints that bedevil humans, have all yielded to the ingestion of adequate daily water. Dr. Batmanghelidj identifies Alzheimer's as due to dehydration of the brain!

Eight 8—ounce glasses of daily water are the recommended regimen advocated by the doctor, to keep the human body fully hydrated. For each cup of coffee or other caffeinated or decaffeinated drink, an additional compensating 8—ounce glass of water is required.

Perhaps the most welcome and convincing finding of the doctor is the key role played by dehydration in creating back ailments in people. The spinal discs consist of about 80% water. When these discs' atrophy through varying degrees of dehydration, an afflicted human is off on one of the most miserable medical merry—go— rounds known to man. The human spine has to support about 75% of body weight. Without fully— hydrated and resilient spinal discs, the spine cannot perform its function properly. "Back problems" ensue. Abnormal physical and nervous pressures develop and vertebrae become misplaced as spinal discs' collapse. In the U.S.A., such back problems are now on an epidemic scale, largely defy physicians and surgeons, and endlessly persecute the afflicted.

Dr. Batmanghelidj has found that rehydration restores the integrity and resilience of the spinal discs. Simple exercises he has devised, create a natural vacuum effect: that draws the needed water back into the discs, whose proper bearing function is thereby eventually normalized. Contrast this with the dead—end, mechanistic "back surgery" that now consumes 1 OOS of millions of $$$ in surgeon's fees and hospital: costs every year.

Back pains are among the body's many cries for water. The horrific wheezing of an asthmatic, which is one of the most embarrassing and distressing things anyone can witness, is similarly the BODY CRYING FOR WATER! Dr. Batmanghelidj has demonstrated that asthma is due to the body's natural histamines constricting the lungs to limit any further loss of water via the breath. The person so afflicted is very desperately dehydrated. In the prison environment in Iran, adequate water provided a cure for asthma one of the astonishing clinical results that first started the doctor thinking about the medicinal powers of water.

Dr. Batmanghelidj considers as " scientific absurdity n which hyper— tension is classically treated. The dehydrated body is desperately trying to hang on to its water volume.

Uncomprehending physicians intervene with diuretics and literally force more water out of the already dehydrated body. The book gives lucid, comprehensible descriptions of the exquisite hydraulic design and engineering of the human body by the doctor; and the diverse functions of the cells, capillaries and membranes as they operate to compensate for dehydration. If the body's many cries for water are ignored, as they are in contemporary America, degenerative diseases are the inevitable consequence.

A political congress scrambling to deal financially with an avalanche of degenerative diseases, is verification enough that the time has come for some new medical paradigms.

Dr. Batmanghelidj has been making a Herculean effort to pass the blessings of his research and works into the lives of all human beings worldwide. He is not permitting his work to be sequestered or short— circuited by very reactionary orthodoxy. He will need all the help we can give him now and in the future. A commentary is appended, to outline some of the biopathic reactions to DR. BATMANGHELIDJ' S LIFE— GIVING DISCOVERIES THAT CAN BE EXPECTED FROM OUR ANTICIVTLIZATION!

FINALLY, I'D LIKE TO RECOMMEND TO OUR READERS
THE EXCELLENT BOOK BY

TREVOR JAMES CONSTABLE ENTITLED "THE COSMIC
PULSE OF LIFE". . .

Trevor James Constable also has an excellent video available for sale pertaining to his works worldwide in 'WEATHER ENGINEERING' /CLOUDBUSTING. He still lives somewhere in Hawaii.

CHEMTRAILS, HAARP AND GEOENGTNEERTNG

Before he was assassinated, President JFK affirmed in a famous speech that "we are opposed, around the world, by a monolithic and ruthless conspiracy". Years earlier, former FBI chief J. Edgar Hoover had stated that "the individual is handicapped by coming face to face with a conspiracy so monstrous h or she cannot believe it exists. 't Before Weccf W i (SOLQ that former President/said that "there is a power somewhere. so complete, so pervasive, that they — people who knew about it — had better not speak above their breath when they speak in condemnation of it. " More recently, Senator Inouye said there "exists a shadowy government with its own Air Force, its own Navy, its own fundraising mechanism, and the ability to pursue its own ideas. free from the law itself. "

This dark secret empire is real and now possesses a weapon so colossal and diabolical that we urgently need to upgrade our ideas of what is possible. We need to put this weapon on our radar and become aware of its existence and operations. To refer to

HAARP (High Frequency Active Auroral Research Programs), a giant field of antennae in Alaska, capable of emitting electromagnetic beams into the electrically charged ionosphere. From there the beams can be reflected downward with very disastrous effects, able to cause earthquakes, seaquakes and tsunamis, volcanoes and damage far greater than a nuclear bomb, without the fallout. A few scientists, including this author, have exposed how the

'ILLUMINATI" now regularly use several 'HAARP' systems worldwide to blackmail nations and deliberately induce earthquakes as occurred in China, Haiti, New Zealand, the pacific in 2004, Haiti and in Japan recently in 2011, and in other countries that don't go along with their plans. The 'HAARP' system was originally developed by the late, great inventor and scientist NTKOLA TESLA, who also developed our A. C. power grid system and many other generating power systems, including the one in Niagara Falls; he also lived and worked in Colorado Springs, Colorado, until he passed away. He was actually poisoned to death by agents of the U.S. government! Unfortunately, this brilliant scientist does not even appear in most text—books in schools, Colleges and many Universities worldwide!
THIS IS A TOTAL DISGRACE TO OUR EDUCATIONAL
SYSTEM HERE TN THE U.S.A. AND
WORLDWIDE! Instead, bird-brain Edison is being mentioned for work in D.C. systems
primarily, TESLA = IGNORED!

From now on, we need to think twice and investigate closely whenever we hear about a so—called 'natural disaster. The HAARP systems are the ultimate weapon of the Globalist criminals, as they have been creating the secrecy surrounding its very destructive capabilities. Also, these systems are very powerful in control of the ionosphere and the uses of these very powerful pulse beam weapons anywhere they use them in their operations. THEY CREATE THE PROBLEMS, CAN CONTROL THE REACTIONS AND THE SOLUTIONS USING THESE VERY
OFFENSIVE WEAPONS AS THEY ARE DOING THEIR BEST TO MILITARIZE OUR TONOSPHERE IN SPACE! The Globalists have plausible deniability and can hide from blame simply because the average person does not know of HAARP I s existence and powers worldwide! For years now the Globalists have been creating numerous 'False Flag Operations' of huge magnitude like the events that happened on 09—11—2001 in New York city,
Washington, D.C. and in   Pennsylvania, along with the 'Gulf Oil' disasters in 2010, and then escaping suspicion while countless people have suffered and died. . . There are telltale signs when HAARP is activated. One is that strange aurora—like lights appear in the sky. Another is that

clouds adopt a very unnatural stratified look. Another sign is that HAARP activity can be measured by an induction magnetometer. All in all, the evidence points to the very strong probability that the recent nuclear Japanese disasters at Fukushima were caused by HAARP, given the seemingly in— explicable and rapid heating of the ionosphere directly above the epicenter of the quake. If you 're still skeptical of all this, please consider what the then U.S. Secretary of Defense William Cohen, admitted back in1997: "SOME OTHERS ARE ENGAGING IN AN ECOTYPE TERRORISM WHEREBY THEY CAN ALTER THE CLIMATE, SET OFF EARTHQUAKES, SEAQUAKES AND TSUNAMIS AND VOLCANOES REMOTELY THROUGH THE USE OF ELECTROMAGNETIC HAARP SYSTEMS WORLDWIDE." HAARP and other electromagnetic weaponry is only part of a larger scheme known as "GEOENGTNEERING", which defined as planetary—scale environmental engineering of the atmosphere + ionosphere, the weather, the oceans and the Earth itself. This is a massive multi—layered operation stretching across all of the continents to achieve very dark objectives. The fact that Congress and the Environmental Protection Agency have been threatened to stay away from INVESTIGATIONS shows the immense money and power behind it! Another very disturbing aspect of geoengineering which is closely related to HAARP is the important phenomena of chemtrails which are trails in the sky left by airplanes after the deliberate spraying of various poisonous heavy metals including aluminum, barium, strontium and various fungi and pathogens. There has been a great awakening by the masses about the existence of these chemtrails worldwide, as people are looking up and noticing that instead of being blue, the sky is crisscrossed with long, white lines hanging in the air for very long periods and eventually breaking up, as you can see for yourselves! The gray, poisonous haze lasts for many hours!

The chemtrail program, known as project Cloverleaf, is very compart— mental i zed; even the pilots have no real ideas of why they're actually spraying. For years this criminal Government has denied the existence of chemtrails, falsely claiming that they were contrails, the water vapor that comes out of jet engines and disappears after 20 seconds. However, the truth is that chemtrails don't come out of the engines, but rather nozzles above the engines, which are fed by tubes inside of the plane. Chemtrails linger in the air for up to 24 hours, after covering the sky, often blocking the sun, and resulting in very un— natural cloud formations. In addition to blanket denials, the Government has been using the excuse and very phony science of this global warming scenario to justify the spraying. GLOBAL WARMING NEVER HAS EXISTED, AS MOTHER EARTH HAS REALLY SHOWN A
COOLING METHODOLOGY IN MANY AREAS WORLDWIDE,
INCLUDING BALANCING WEATHER CONDITIONS EXTENSIVELY AS A LIVING, BREATHING + RENEWING COSMIC PULSE OF LIFE PERVADES IT!

What are the aims of chemtrails? There are several. Firstly, similar to other depopulation plans such as the use of poisonous vaccines, fluoride, aspartame, the objectives are to gradually poison the use— less masses, but to do it so slowly that no one can discern the true cause. People find it harder to heal their diseases if they don't know the origin of their illnesses. Aerosol spraying is particularly now effective since everyone has to breathe, and once it's sprayed over— head in many areas, people can't easily escape and are essentially forced to ingest the toxic particles. The idea is to make people sick, infertile and trapped in survival mode, where they are so focused on just getting by that they have little or no energy to investigate what is happening and mount an effective resistance.

Secondly, large biotech corporations profit from the spraying. When the chemtrails fall to earth, they land in the soil and raise its pH, making it too alkaline and unable to support crops. Monsanto is developing aluminum resistant crops so they can monopolize the market, force organic crops out and get everyone to buy their 'Genetically Modified' seeds. These 'terminator' seeds produce no further seeds when planted, so farmers have to keep buying seeds from Monsanto every year.

Thirdly, HAARP and chemtrails are working together in a program of iron fertilization ', in which the planes specifically dump millions of tons of barium and iron into the seas, which changes their com— position, HAARP then magnetizes these metals in the water to create high and low pressure systems, allowing it to control the weather and create storms and floods at will! Likewise, other particles found in chemtrail cocktails may be enabling HAARP to manipulate our thoughts by energizing the otherwise nonconductive air we breathe, so HAARP is a mind control weapon too.

Fourthly, there is a very huge amount of activity currently happening with our Sun, especially large solar flares, as I had previously indicated! Constantly, there are great changes occurring on our planet and in our solar system- and galaxy which have the potential to then activate our DNA and propel us forward into our evolution to dimension #5. The Controllers fear a global awakening and are desperate to keep their grip on the plane, so they 're dumping these toxic metal particles in the stratosphere to act as a shield, block the sun's rays and obstruct the activation of our DNA.

Fifthly, the metals which linger in the stratosphere provide a better compositional backdrop for the projection of holograms. You may know of Project Bluebeam, the plan to fake an alien invasion or 'second coming' of Christ by projecting a giant and very life— like hologram into the sky. The idea is that, in a situation like this, people would be so scared or excited that they would willingly give up their power to a world government claiming to guide or protect them! SO,
NOW YOU CAN PREPARE YOURSELVES FOR THE MANY
MAN-MADE DESTRUCTIVE TECHNOLOGIES THAT ARE TN
OUR VERY IMMEDIATE FUTURE, LADIES AND GENTLEMEN
EVERYWHERE! MAY GOD BE WITH YOU ALL, ALWAYS IN
THESE VERY INTERESTING "END TIMES" FROM NOW UNTIL THE COMING EARTH CHANGES END
TN YEAR 2060! THANK YOU!
NATURAL WATER RESOURCES ARE MAN-MADE

The book "Pillar of Sand" is a book that you don't need to read — but you want to beware: The book is a product of the international propaganda mill, 'The Worldwatch Institute', for which the author, Sandra Postel, is now a senior fellow, and was formerly vice president for research, from 1988—1994. She now heads the Global Water Policy in Amherst, Massachusetts.

Postel reviews locations around the globe where there are many water shortages, and then she attributes water resource depletion to over— use of technology, in particular, technology for irrigation. Moreover, she includes a chapter on the history of irrigated agriculture in ancient settings, concluding, "History tells us that most irrigation— based societies fail." Well, that's just utter nonsense, due to the fact that Mesopotamia and Egypt had many wonderful irrigation canals that had been built very effectively, going back to 4,000 BEFORE PRESENT!

In reality, under the very bad and worsening economic conditions in almost all nations over the past 45+ years, advanced science and technology have been held back in infrastructure, industry, agriculture, medicine, and other areas. National economies worldwide are failing because of the 'under—use' of technology, interlinked with the spread of " free " rigged trade, too many speculative financial bubbles, and global ism, backed by the very political and financial interests Worldwatch serves the lousy International Monetary Fund, the World Bank, and many related private circles.

There are many examples. In Africa, there has been no modern, high— tech river—basin development, nor nuclear—powered desalination, which would allow fabulous irrigation, and agricultural and industrial improvements. In North America, nuclear— powered desalting systems have not been built as originally planned; for example, the abandoned plans for desalting in southern California and on the Atlantic coast, where saltwater incursions are severe. Instead, more destructive nuclear power plants are being built all over the United States. More nuclear disasters are going to occur as the result of the criminally stupid politicians having to spend billions of $$$ on these major catastrophes

about to happen from now through the years 2060! THE U.S.A. IS NOW TOTALLY BANKRUPT + THESE STUPID
POLITICIANS WILL HAVE 100 TRILLION $$$ TO WASTE FOR
THEIR INSANE WARS DUE TO MAINTENANCE COSTS FOR THEIR MILITARY INDUSTRIAL COMPLEXES WORLDWIDE! How many trillions of bankrupt U.S. $$$ have already been spent on the black projects that are not being reported since 1946—1947, when these black projects were started? Anyway, the NUMEROUS EARTH CHANGES WILL DEFINITELY HAPPEN, AS THEY'VE STARTED BACK IN THE YEAR 2004! Numerous 9.0 - 9.9 earthquakes will strike down MILLIONS IN THESE COMING EARTH CHANGES, SO BE PREPARED FOR THE ULTIMATE ARMAGEDDON WORLDWIDE!

Returning to the water scenario, 'water expert' Postel does not criticize such advanced technologies as nuclear—powered desalination techniques, nor electron— beam— treatment of wastewater, nor other high— tech water supply systems: She does not even mention them! Nor does she even identifies advanced water—usage systems in agriculture — hydro— ponics and aquaculture, for example.

Instead, she proposes a list of practices, many of them low—technology, which she refers to as her "Blue Revolution" strategies. For poor countries, this means foot— treadle pumps for lifting groundwater, for low— cost bucket—drip irrigation. For California, this means relining canals, recycling farm run—off, and so on.

She also includes a Blue Revolution proposal for "water markets." This exactly complies with the demands of global commodity cartel interests, best known as the "BAC" British—American— Commonwea1th grouping: to monetize and privatize water " rights," along with all other vital economic commodities. Postel writes: "Until fairly recently in the U.S. a variety of federal and state laws and regulations inhibited farmers from selling their water. As legislatures and courts gradually sweep away these restrictions, markets are opening up.   WORLDWATCH'S PLAN: CUT PEOPLE

The fundamental premise of Worldwatch, founded in 1974, is that population growth has exceeded science and technologies. This is the outlook of Worldwatch's founding sponsors among the international financial and political circles, who want power and control through global ism. The key foundations providing financial backing reads like a "Who's Who" of heirs to the British East India Company, including the Rockefeller brothers, Pew, MacArthur, and other leading foundations.

Postel, a Pew Fellow, was an adviser to the Global 2000 anti—population growth program, and is a member of the board of the World Future Society and consultant to the United Nations Development Programme and World Bank. Her writings on water questions are entirely ideological: She presents a series of assertions and sophistries. For example, she states that the greenhouse effect will raise temperatures, which will affect snowpack, which can cause flooding, which will then cause more "risk" to irrigation systems.

The message of Worldwatch is that its every dismal utterance is heavily promoted, and accorded credibility. Postel's 1992 book on water, Last Oasis, appeared in 8 languages, and was used as the basis for a Public Broadcasting System documentary in 1997.

The poet Shelley is warning us of the danger of tyrannical blockheads, not "technology" and progress only if we allow the arrogance and stupidity of today's arrogant colossi the IMF, the BAC, the major media, and the mega think—tanks like Worldwatch — to prevail, will it mean sure death. Listen to the poet, and the future is beautiful.

GREAT PROJECTS

The poetic truth is borne out in the array of technologies and projects that can be used to "make" natural water resources. Let's look at 2 categories as examples geographic engineering and advanced energy water production: On every continent there are large— scale water development projects unfinished, or never begun.

In North America, there is the NAWAPA, the North
American Water and Power Alliance, designed in the 1950s, reviewed by Congress in the
1960s, and then shelved after the post—industrial shift in the 1970s.

NAWAPA would redirect southward, about 15% of the Mackenzie River flow t is now going toward the Arctic, and utilize the 500—mile wonder of the British Columbia Rocky Mountain Trench, channeling the water through a network serving the Canadian Prairies, western United States, and, indirectly, or directly, Mexico.

The hydraulic designs for the Sierra Madre, developed by Mexico 's College of Engineers 40 years ago, are sister projects to NAWAPA, called "Hydraulic Project for the Northwest" and the "Hydraulic Project for the Gulf of the Northeast". Via NAWAPA, U.S. water supplies would be increased by 20% — 135 billion gallons a day.   In Africa, the centerpiece project is the 1970s Trans— Aqua plan, to develop the Zaire River Basin, with inland lakes, ports, hydro—power, rail, and highways.

In the Middle—East and other arid regions, nuclear— powered desalinating plants, strategically located on a Mediterranean—Dead Sea ("Med—Dead") canal route, along the Suez, the Gulf of Aqaba, and Red Sea, would transform the region, with power and water supplies. Just 20 of these facilities would equal a "New Jordan River" in volume!

On the vast Eurasian expanse, the priorities include resuming the southward diversion of parts of the run— off now flowing toward the Arctic, to provide fresh input into the Aral Sea Basin. Work on these diversion canals was started 35 years ago, and then suspended. There are priority projects for China, the Mekong Basin, Australia, and for the Southeast Asian archipelago nations, where nuclear—powered desalination facilities can mitigate the terrible episodes of El Nino drought.

DESALTING SEAWATER

Desalting seawater requires reducing the parts per million (ppm) of dissolved solids (80% of which is salt) from 35,000 ppm to less than 500 ppm, a reduction of 70 to 1.

There are several methods now commonly used: distillation (some form of which is used in more than 90% of installed desalination capacity), reverse osmosis membrane (newly improved), electrolysis, and vapor compression. In addition, research into the electromagnetic structure of water promises revolutionary methods of desalinating for the future.

With the many recent advances in materials involved in seawater de— salting, the chief cost of making fresh water is the energy involved. Using advanced nuclear generation to provide power inexpensively, greatly reduces the cost of desalination, for example, the gas— turbine modular helium reactor (GT—MHR), proposed by General Atomics, based in San Diego, can be efficiently used for both energy generation and water desalination.

The application of these new technologies would provide relatively low—cost water, along with electricity, and create new " run—off" at strategic coastal sites; in other words, we would create new supplies of water that are equivalent to new man— made rivers and reservoirs.

One proposed installation of the GT—MHR, and a desalination facility (multi—effect distillation) in Southern California was projected to provide 106 million gallons per day, which is comparable in size to Atlanta, Georgia's municipal water system (104 million gallons per day, serving 700,000 people), and that of many other cities, including San Diego (104 mgd, 1 million people) and Honolulu (110 mgd, 700,000 people).

Just a brief look at some of the new technologies that could be available makes a sham of Sandra PosEe1,
Worldwatcher and their gloomy prognosis for mankind.

The author of this data was Economics Editor of the weekly Executive Intelligence Review. Her review of "Mideast 'Mega—projects' to Build Infrastructure and Peace,' which centered on economist Lyndon LaRouche' "Oasis Plan" for building nuplexes nuclear—powered agro—industrial centers appeared in the spring   1995 issue of 21st Century. Her name was Marcia Merry

            Baker.

THE CURRENT "BTG BANG THEORY IS DEAD WRONG!

This ewritten transcription of a theory that was written years ago by the scientist &a retired nuclear engineer and also a physicist, residing now in Southern California county. This data is private and I'm not going to release any other information pertaining to this scientist. Some of this information has also been researched by this author, dear readers at large. If you need to contact me, you can call me (760) 327-4761, as I DO NOT HAVE ANY "E-MAIL" FOR FREE INFORMATION AVAILABLE. YOU CAN CONTACT THE WEBSITE www.mullerpower.com for data available on the late Bill Muller's magnetic generator, with many different exhibits for your pleasure and enlightenment.

From the viewpoint of modern cosmology, there was supposedly at first, only 1 main event. 15 billion years ago, a 'major black hole' exploded, CREATING THE UNIVERSE??? The universe was supposedly given one initial burst of energy, and it has been winding down ever since. But the 'Big Bang' isn't something we can see through a telescope.

THE BIG BANG IS ONLY A THEORY.

The Big Bang is part of different theories. Tt was invented to explain how the expanding Universe started expanding. "Why are the galaxies moving away from each other? "
In the 1960s, quasars were discovered. Several quasars were already known but not recognized as anything special. They were thought to be stars in our Milky Way with a few odd characteristics, such as their blue—violet color and their association with strong radio sources. Then their redshift was measured. It was much higher than that of the very farthest known galaxies. WHAT A SHOCK!

The redshift yard—stick demands that these objects are far beyond the galaxies. If that's true, how bright must they be? There must exist a problem here for astronomers. An unknown energy mechanism has to pro— duce such intrinsically high—luminosity objects, enabling them to be so bright at such great distances. There was no known mechanism to produce fantastic amounts of that much energy.

SO WHAT REALLY DID OCCUR? We have to notice that many of the newly discovered high redshift quasars seemed to be surprisingly close to the galaxies numbered 100 through 163 in the HALTON ARP catalog. These galaxies are the Seyfert galaxies also called active galaxies and the starburst galaxies.

Many of the quasars occur in pairs and lines and arcs, with a low— redshift Seyfert galaxy sitting nearby. Often the Seyfert galaxy is positioned so that its family of quasars seems to have been ejected from both ends of the Seyfert spin axis. X—ray jets and radio lobes of the galaxy point directly at the line of quasars, often enveloping them. How could this be if the quasars are a universe away from the Seyfert?

1 or 2 or even a dozen quasars positioned near ordinary galaxies might be coincidence. But Seyferts are rare and spectacularly different from ordinary galaxies. They have enormous bright nuclei. Often they' re bracketed by twin giant lobes of radio and x—ray material pouring out in the same direction as their families of quasars. Some, like M82, are disrupted, exploding, torn apart. Others, like M87 and CentaurusA, have jets which stretch thousands of light—years. And from the end of these spectacular jets, these galaxies are spraying quasars across the sky.

What does this mean? It means that, quasars are neither stars or super luminous galaxy cores. Astronomers have replaced one wrong idea with another wrong idea. When restored to their rightful distance, alongside the galaxies they 're ejected from, they become brighter than stars, but fainter than most galaxies. This is newly created matter, which will eventually become a full—sized galaxy.

This means that quasars and active galaxies belong together, in spite of their different redshifts, the distance is the same. This implies that the yardstick of redshift distance' is mistaken. IF THE YARD— STICK IS
WRONG, THE THEORIES BASED ON TT ARE WRONG,
INCLUDING THE BIG BANG! That means it's time to weed out the distortions caused by arranging the universe according to this faulty yardstick. Even the stars in our Milky Way show small non—velocity redshifts.

IN OUR COSMOLOGY, A LARGE COMPONENT OF
REDSHIFT MEANS AGE, NOT VELOCITY.   The higher the redshift, the younger the galaxy or quasar. Galaxies that have no excess redshift are the same age as the Milky Way. The 7 galaxies out of millions which are blueshifted are older than the Milky Way. 6 of these blueshifted galaxies are in the Virgo Cluster and the seventh is M31 also known as the Great Nebula of Andromeda, our nearest neighbor in space and the dominant galaxy of 20—30 galaxies.

SO HOW WAS THE UNIVERSE BORN? According to the Big
Bang theory, the universe exploded into existence about 15 billion years ago, In our different scenario, the same data points to a different event about 15 billion years ago the 'birth' or ejection of the Milky Way.

In our different universe, M31 is the parent galaxy of our Milky Way. 15 billion years ago, M31 was an active galaxy and the Milky Way was a knot of plasma in M31 's jet. Unlike the slanted jet of CentaurusA,

M31 's jet was aligned straight down the spin axis. How do we know this? M31 isn't active today it no longer has a jet. But we can deduce the direction of M31 's jet from looking at M31 's family, our local group of galaxies. After billions of years, this family is still lined up in a remarkably straight line.

The Milky Way is a parent as well. The Magellanic Clouds, visible from the Southern hemisphere, are 2 of its offspring.

So what happens when galaxies grow old? We don't really have enough information yet to know what happens when galaxies grow old. Perhaps they exhaust themselves and fade away.

HOW BIG IS THE UNIVERSE? According to the 'Big Bang, the universe atmosphere about 30 billion light years in radius. In our different universe, all we can know is that it stretches out in every direction farther than we can see. It is also older than we know, possibly infinite and eternal. But for that part in the universe we see in our telescopes, high redshift objects are closer than the redshift yard— stick indicates.

Many modern astronomical concepts, such as curved space—time, are really only attempts to compensate for the distortion of measuring the universe with a broken ruler. Basically, the universe is flat and Euclidean. Gone is the logic—defying curved space which so many people have strained to imagine, not to mention curved time. Gone are the hypothetical singularities — black holes where physics just breaks down, and gone, too, is a universe made up of more than 90% invisible 'matter '

HOW DO GALAXIES EVOLVE? According to the 'Big Bang', scraps of matter from the initial explosion fall together under the influence of gravity to make a galaxy — which does not make any sense! IN OUR UNIVERSE, active galaxies eject new matter in the form of high redshift quasars. The quasars gain mass and lose redshift in even jumps as they grow into mature galaxies. We can know the process today. If you know what to look for you can see it taking place yourself. Galaxies don't "blow bubbles" they eject quasars.
BLACK HOLES DON'T COME IN STANDARD SIZES, THEY DON'T EVEN

EXIST! BLACK HOLES, WHERE EVERYTHING IS SUPPOSED   TO BE FALLING IN, are a very poor explanation for the cores of active galaxies, where everything appears to be falling out.

HOW DOES THE UNIVERSE END? According to the 'Big Bang', there are 3 possibilities: It might continue to expand forever. Or it might reach equilibrium and remain stable forever. Or the expansion might stop and all the galaxies would collapse back into a black hole. NONE OF THESE SCENARIOS MAKE ANY SENSE ...We really have no idea how it will end or if it ever will. THAT'S AN OPEN QUESTION FOR FUTURE RESEARCHERS TO SOLVE HOPEFULLY.

When the distortions of the redshift yardstick are removed, a big picture of the form of the universe comes into focus. The whole population of the sky becomes 2 enormous spiral superclusters. Our local group lies between them, possibly along the arms of the brightest supercluster. These superclusters are centered on the Virgo Cluster in one part of our sky and the Fornax Cluster on the other. Spirals of spirals. Galaxies of galaxies.
AND WHO KNOWS WHAT LIES BEYOND, WAITING TO BE
DISCOVERED? THIS UNIVERSE WILL NEVER EVER END, AS

GOD'S INFINITE INTERDIMENSIONAL ENERGY SOURCES WILL LAST FOREVER WHAT'S THE LATEST DATA CONCERNING OUR NEIGHBOR MARS?

MONOLITHS THE 2 FACES ON MARS AND THE 7 CITIES + LOTS OF METHANE-ON-MARS!

In the 1960s the movie '2001 ' a Space Odyssey, a mysterious monolith suddenly appears on Earth and diverts the course of human evolution. Mankind ultimately sets out for the stars and finds similar monoliths placed along the way on a few of the other planets in our solar system. Does this provide any insights towards 'life' on other planets??? Not surprisingly, 2 monoliths have been photographed on Mars. . . Snapped from 165 miles away by the Mars Reconnaissance Orbiter, 2 rectangular objects can be seen standing on the barren plain. Scientists at the University of Arizona who took the image estimate that the 2 objects are about 5 and 10 meters across. The picture, when posted on the web, caused a sensation among 'space junkies'. • •

NASA, which stands for 'NEVER A STRAIGHT ANSWER', has continued to deny the presence of any evidence indicating intelligent life anywhere in the universe except on Earth, which I call now the planet of the 'human apes' and previously genetically engineered   'humans'.

NASA has not been helped in such efforts by recent public statements from ex— astronauts and moon walkers, Edgar Mitchell and Buzz Aldrin. Mitchell has declared that aliens exist. Aldrin has publicly alluded to a "monolith" seen on Martian moon Phobos and said that we should explore the Martian moons.

It's all simulacra, say 'NASA's experts'. Simulacra is the phenomena which causes people to see familiar images in random surroundings (i.e., animals in the clouds). That was the official explanation of the so— called "face on Mars" first photographed by the Viking spacecraft in the 1960s. The agency's credibility is greatly weakened on the subject — however, by its admitted efforts to distort with hi—pass filters the publicly released hi— resolution photos of the "Face" taken by the Malin Space camera in the 90s. If an object is really the result of natural, rather than intelligent, causes, then close observation should reveal the truth. The fact that NASA chose to interfere with the re— lease of the image of the "face" yielding what has been called the "catbox" image could be plausibly interpreted to mean that NASA had something it wanted to hide! SO, WHAT ABOUT THE OTHER FACE#2 ON MARS???

Face #2 on Mars is located in the south polar region of
Mars. Please obtain a copy of my previous book entitled "GRAVITY, MATTER & SPACE TRAVEL", published in 2006. It clearly shows the face of a woman that can only be seen from space. It is about 1 mile in diameter, as you can see for yourselves. It also shows photos of 'sand—whales t on the watery surfaces of Mars.
We are missing something very significant about Mars and one of its most important signatures which could indicate life. That is the opinion now expressed in the journal 'Nature', of French scientists Franck Lefevre and Francois Forget of the Université Pierre et Marie Curie in Paris. They're talking about methane on Mars.

Methane can be created by volcanic activity or by living organisms. Whether it comes from biology or geology, though, there is a lot of methane on Mars, coming from somewhere, but it is also destroyed very quickly. Lefevre and Forget say the surface environment on Mars is very hostile to organics. That much is inferred by the rapid rate at which the methane is destroyed there. That does not mean, however, that conditions beneath the surface may not be-more benign. In any event, something on Mars is creating lots of methane right now in a seasonal cycle. The chemistry of the Martian atmosphere, however, remains a mystery, quite unlike anything we observe on Earth. "There is something else going on," say the scientists, " that lowers the methane lifetime by a factor of 600. So if the measurements are correct, we must be missing something quite important."

The presence of large amounts of methane on Mars has been detected by telescope through a process called infrared spectroscopy. According to Dr. Michael Mumma, director of NASA's Goddard Center for Astro— biology, 19,000 tons were measured in a plume in 2003.

If the Martian methane is the result of organic life below the surface, it does not seem unreasonable to speculate that there may be quite a bit and that it may be still living.

It is important for readers to obtain copies of my previous book, as mentioned to study in much greater detail what is really going on in the 7 underground cities on Mars. Furthermore, time travel has already been achieved years ago to Mars, as is known by a few researchers

— including this scientist! NASA's lies only add to fact that life does exist on Mars, as has been shown by this scientist & other researchers!

A HEALTHY MIND TN A HEALTHY BODY

It is an overused phrase, but it has lost none of its truth.

To achieve the perfection that will open the doors of all possibilities to you, meditation will be useful and necessary, but it won't be enough. It has to be supplemented with 2 marvelous activities which will over— haul your life in the most positive way possible:

YOGA AND TAICHI

Yoga

Yoga is a method of overall development. It includes several branches which contribute to the blossoming of the human being at different levels: physical, psychosomatic, mental and spiritual.

Yoga has been part of ancient human knowledge for thousands of years and is of Indo— European origin. It was passed down to us by Indian and Tibetan yogis, and has been an integral part of man's heritage ever since.

There are many schools of yoga and numerous methods throughout the world. Yoga operates by allowing men and women to manage their own health and personal development by physical, psychosomatic and mental means. It also involves techniques of hygiene, nutrition and dietary supplementation. Its aim is to distance man from suffering and illness by a new self— consciousness, but particularly to let him take charge of himself overall.

Beyond its contribution to good health, the aim of yoga is to give everyone a practical means of self— development and human accomplishment. Although it must be taught by/qualified and experienced person, yoga remains a method based on personal effort, freedom and individual ability. Whoever practices it takes the reins of his future into his own hands and acquires the means of real autonomy.

Where does yoga come from?

Yoga originated in India. It has its roots in ancient Indo—
European civilization. One of the oldest works on yoga (Patanjali's yoga sutras) is dated about 500 B.C.: it is a work treating Raja yoga.

In what order should yoga be studied?

Traditionally yoga moves up in steps from Hatha yoga to Raja yoga. It is therefore preferable to study the basic discipline, which is Hatha yoga, first.

This consists of the development of the body and breathing, as well as the ability to concentrate and relax (first level yoga Nidra). Although they take several months to learn properly, Hatha yoga techniques give good results very quickly when they are correctly performed. Once Hatha yoga has been acquired, the study of Kundalåni yoga and Raja yoga can begin, if desired, as well as the study of second level yoga Nidra techniques.

Yoga is a quest which deepens over the years because it is based on an awareness which develops constantly with practice.

Raja yoga opens a limitless field of experimentation and exploration of human awareness. Yoga is a critical element for success in your life, and to make the effect complete, discover the joy of practicing Tai Chi!

Tai Chi

TAICHI CHUAN (or Taiji Quan) forms part of the Nei Chia (internal martial arts) of Chinese origin. These origins have been lost in the mists of time.

These arts are considered soft, in contrast to the Wai Chia (external martial arts), which are considered hard, for example Kung F u. They are also known as shadow boxing, supreme heights or grand reversal boxing, the dance of life, eight trigrams boxing (Y i king) etc. According to its schools and styles, Tai Chi has multiple aspects: Tai Chi form practice

This is fundamental, whether alone or in a group. It is the base for all the work which follows and suits most practitioners because of its relationship with relaxation, health and meditation.

Push

(Tuishou, Sanshou) with a partner is very playful and enriching. It enables you to try out, better understand and apply the form movements in their more energetic and martial aspect.

Weapon forms

Whether sword, sabre, fan, stick or pole, weapons are an elongation of the body, whose mastery favours your own self—mastery. As you will understand from its description, this discipline represents a fabulous complement to meditation and yoga for you.

It is this package which will enable you to change your way of looking at the world and at life, let you develop freely and without impediments, with an understanding and perceptiveness that is bound to lead to your success. It's one of the most fabled quests in the history of all mankind:

THE SEARCH FOR THE FOUNTAIN OF YOUTH

From the very beginning of the written word in ancient Mesopotamia, tales and legends have been written about foods, potions, medicines, herbs or special waters that could stop the aging process and keep people forever young.

In 1493, on the second voyage of Columbus, a Spanish explorer by the name of Juan Ponce de Leon accompanied Columbus to the New World.

Ponce de Leon heard tales from the local Indians of an island called Bimini, located somewhere north of Cuba, which reputedly possessed the fountain of youth, a spring whose waters had the power to restore youth.

Even in the 15th century, the promise of renewed youth was a powerful motivator to what were then the world's greatest explorers. The search for this Fountain of Youth led Ponce de Leon to discover Florida where he continued his search _for the miraculous source of youth until he was mortally wounded in battles with Indians. Ponce de Leon died trying to find what millions of others have searched for thousands of years a way to stay young, a way to never grow old.

The dramatic story of Ponce de Leon represents a fundamental drive of the human race. That drive is to not only survive as a species, but to survive as individuals, and not just as individuals — but individuals who are young, healthy, vigorous and engaged in life for as long as possible.

After thousands of years of civilization, mankind may finally be coming close to one of its fondest dreams the true attainment of long—term youth.

As it turns out, the true Fountain of Youth is not something out there in some mysterious far off land, just waiting to be discovered by some heroic explorer.

NO, THE FOUNTAIN OF YOUTH IS RIGHT INSIDE OURSELVES, AND HAS BEEN THERE ALL ALONG! WE CAN MAKE THIS CLAIM ON 3 LEVELS:

FIRST, maintaining a youthful body, mind and lifestyle seems to be intimately connected to a few hormones which are produced naturally by our own human bodies. By monitoring the levels of those hormones and maintaining them in a balance that is appropriate for a youthful body.

THE JUPITER NUKE AFFAIR

Stranger to neither scientific controversy, Richard Hoagland used his Enterprise Mission site to advance the stunning hypothesis that in order to prevent contaminating potential life on Jupiter's much studied mccn, Europa, NASA chose to deliberately crash the Galileo probe into Jupiter, and in doing so, NASA inadvertently caused a nuclear explosion when the on board a radioisotope thermoelectric generators, hereafter RTGs, used to power instruments, implod dunder enormous pressures and temperatures as they descended into the Jovian atmosphere.

Evidence presented includes observation of the planet before and after the impact, U.S. nuclear tests, Jovian atmospheric models, and the engineering and design of the RTGs themselves. The full details are provided at the following link: http://www.enterprisemission.com/NukingJupiter. html

Readers should consider the purported Jovian nuclear detonation as being merely a point of departure for what follows. Prepare, then, for a trip not just through Alice's looking glass, but into realms far beyond. "What's of great interest and extremely significant is that in the acknowledgements of *Clarks 2010*, he alludes to a communication he got from NASAs Dr. Walter Jastrow who openly admitted that his *Lucifer Thesis* was of great interest to the Agency – with regards to the Galileo Mission which was then a proposed exploration of Jupiter. "

This brilliant concept may provide one or the best motives for the projected Galileo Mission. 'Taken by itself, such a comment seems a trifle odd, but it assumes an entirely different set of intensions when viewed against esoteric practice, subsequent events, and even thorough analysis.

The Strange Space
Program
Millions of American taxpayer's dollars are being fed into it.

Unbelievably yet somehow, some believe something else was first piggy—backed upon the overt space program, then later came to dominate it. Instead of an agency accountable to the people, we now have, and have repeatedly seen, an agency which behaves as a law unto itself. NASA's handling of the reimaging of Cydonia, is a textbook example.

NASA bitterly fought demands for new, high quality Cydonåa images for years, before being forced to comply.
According to astronomer Tom Van Flandern and others, NASA did everything it could to minimize both the scientific take of that effort and to marginalize the findings via clever selection of suboptimal imaging angles, bad lighting, etc. Further, having severely crimped what got imaged, it then sought to bury the take through a clever exclusive analysis deal with Dr. Malin and his Malin Space Science Systems. When public pressure to honor its initial agreement forced the release of at least one image, NASA then removed most of the information content of a blockbuster image of "the Face" on Cydonia by putting it through a high pass filter, thus ensuring that the now infamous

"Cat box" image was what made the evening news cycle and was imprinted on minds worldwide. Only after the deed was done did NASA grudgingly release a proper image of the face. By then, of course, the story was forgotten "old news. That was bad enough, but it came amid a pattern of repeatedly occurring launches coinciding with certain ritually significant stars and stellar positions. Spacecraft said lost by NASA were, according to insiders, still fully operational, with their take kept hidden from the public.

Later experiments were conducted on days when NASA said none were. Blatant digital manipulation distorted released images. Even IV Åis— plays at JPL, it is alleged, were deliberately reset in order to sell a particular concept of Mars.

NASA's/JPL's peculiar heritage is revealed, it is charged, by selection as head of Pasadena's Jet Propulsion Laboratory of "Jack" Parsons. Parsons, it turns out, was the notorious Aleister Crowley's handpicked lodge master for the Pasadena, California chapter of the occult, magical organization 'Order of the Eastern Temple'. According to Hoagland's associate Nick Skouras, writing in his research paper JPL, Parsons, and the Crowley Connection, "Parsons swore an oath of allegiance to guide humanity into 'an age of communion with the Gods, using his vast arsenal of talents, both scientific and magical. " Hethen goes on to say "From the pen of both Crowley and Parsons, their goal was to continue this practice beyond this earth, extending into the great temple of space in order to, once and for all, make 'communion with the Gods' a reality." Elsewhere, Nick Skouras indicates in his paper that these "Gods" are extraterrestrials, but the writer believes this may be too narrow a viewpoint, since it excludes many other possibilities. What is clear, though, is that over and over again, we see unusual relationships between NASA missions to the Moon, Mars, and elsewhere and the constellation Orion and the elevations or depressions of it 19.5 or 33 degrees from the horizon. These numbers are of great occult significance numerologically, and go to such matters as genetics and hyper—dimensional physics as well to those realms covered by Arthur C. Clarke's remark "Any sufficiently advanced technology is indistinguishable from magic.

SECRET WORSHIP

In his many books, Manly P. Hall tells his readers repeatedly that, since ancient times, the gods were worshiped in 2 forms: exoteric and esoteric. The first is the sort of thing open to all, is conducted by rote, and performed with little or no understanding by the "faithful t' of what's really going on. Were it milk, it would be low fat, whereas the esoteric is the Hidden Knowledge, a ritual and form of worship known only to, and practiced by, a handful of carefully chosen, sworn to silence initiates. Were it milk, it would be whole milk, straight out of the cow, cream intact. See the difference?

With the above in mind, let's see what we can make of this most un— governmental interest in Orion. Orion and its worship under different names figure prominently in many occult traditions and are attested to in archaeology and ancient literature. Orion, is, in turn, but one of many names for Osiris. Osiris is also said, in this context, to be another name for Lucifer.

How does this help us with our nuking Jupiter scenario? Simple, NASA's peculiar launch practices are, in part at least some have argued, tied to Orion/Osiris worship, or as some suggest Lucifer worship.

Investigation of these repeated "coincidences" by a trained statistician concluded that these were coincidences only if one were com— for table with odds of a trillion to one against.

COOPER's TALE

The late Bill Cooper's life, as presented in his astounding 'Behold A Pale Horse ', was both strange and interesting before 1970 when he joined the Intelligence Briefing Team of Admiral Bernard Clarey, Commander in Chief
Pacific Fleet, but it's what he learned after he got there which concerns us here. specifically, his separately claimed encounter with Project GALILEO in 1972 5 full years before what became Project Galileo was conceived, per NASA's official Galileo site, and 10 years before Clarke penned 2010. And what did Cooper say he saw? 'The unbelievable pressure that will be encountered will cause a reaction exactly as occurs when an atomic bomb is exploded by an implosion detonator. The plutonium (a whopping 49.7 pounds in the RTGs) will explode in an atomic reaction, lighting the hydrogen and helium atmosphere of Jupiter and resulting in a star that has already been named LUCIFER. The world will interpret it as a sign of tremendous religious significance. It will fulfill prophecy.

On that same page 72 of his book, Cooper calls it "the insane application of technology by the JASON society which may or may not even work, 't and goes on to note "documents that I read while in Naval Intelligence stated that Project GALILEO required only 5 pounds of plutonium to ignite Jupiter and possibly stave off the coming ice age. Global warming is a hoax. " The planned date of the creation of the new, baleful sun? Midnight of December 31, 1999 at Giza, Egypt — with the global power and occult elite in attendance at closed ceremonies gazing knowingly at the heavens, waiting eagerly for the blaze from their engineered Luciferåan sun! Happily, neither the planned detonation occurred nor the planned installation (aborted by outraged Muslims) of a gold capstone (Lucifer again?) atop the Great Pyramid.   Initiates in policy—making positions in NASA, it is alleged, have for decades practiced covert ritual worship in the heavens of their "Gods' at taxpayer expense. Arthur C. Clarke, it's suggested, could be: a) an occult initiate; b) primed by such initiates, or c) may have simply tapped into the then unmanifest. In any case, elements within NASA clearly resonated with his "Lucifer Thesis." Equally, the supposed "fuel" load for the RTGs was ridiculously large, so much so that quite a few people openly worried that a nuclear disaster might ensue should Galileo fail to launch properly.

If we take Bill Cooper at his word, we also have a separate, earlier confirmation of Hoagland's implosion scenario, together with the reasons (engineered announcement in the heavens of Lucifer's New Age and a way to perhaps save earth from another ice age) He also tells us that the planned device and detonation method were of such dubious reliability that those behind the scheme hedged by increasing the fissile material by a factor of almost ten. Even so, if Hoagland's right, the plan fizzled. There was no " second sun" at the turn of the century, and the purported nuclear event from the deliberate crashing of Galileo into Jupiter's atmosphere on September 21, 2003, while it stirred things up by October 19, 2003, certainly brought forth no end to night on earth and yes, September 21 has ritual meaning. See link http://www.goroadachi.com/etemenanki/archive— 2003d.htm (under "September 22" for details.) ATLANTIS, SCALAR WAVE WARFARE AND THE ETERNAL STRUGGLE BETWEEN GOOD AND EVIL What if someone uses Systems technology for selfish and evil purposes? This has already been done. 12,500 years ago the Atlanteans did just that. During the peak of Atlantean times, the DNA was so well—tuned that mankind had an average lifespan of over 800 years,

Intellect flourished and priesthood was saintly. But as we reached the backside of the equinox, the Dark Forces intervened and great temptation reigned. The Grays appeared and Scalar Wave Wars erupted, causing great destruction.

Because of the misuse of scalar waves, the DNA of humans was affected and lifespans decreased by 90%. The offensive, destructive force released carved huge negative vortexes into the cities that were toppled as the civilization and science of those times became less than history almost mythical to the average human brain with its severely limited genetic memories! The famed Bermuda Triangle is One of 3 such spots still acting as an astral doorway when planetary conditions amplify lunar energies. Poseidon was the principal city of Atlantis. Here were located the ancient mystery schools, spaceports and Hall of Records for Akasha. This bubble of positive existence became a black hole of negativity much like the fall of the Roman empire after the demise of Mark Anthony and Julius Caesar. Strangely enough, the island of Poseidon was located exactly 3 miles off the coast of Florida near Miami, right in the heart of the Bermuda Triangle.

Today Miami is still thought of as the drug and crime capital of the world, and as Earth's aura retains more recently incarnated souls, other cities located in other vortexes are fast becoming full of recently executed criminals that have reincarnated back to regain their vengeance and destructive ideals until humanity wakes up and GETS THE BIG PICTURE!

Let us examine a few details for the Big Picture.

First of all, let us consider the Universe as a precision Swiss watch. We're beings living somewhere inside of this watch, tiny little specks residing like a little gear deep within the highly jeweled,

precision works. As the watch ticks away and gears mesh, time is measured by the scale of numerals numbered 1— 12 with a big hand for one point, it tells us that the sons of Saga ras went off in the north— eastern direction from India and entered into the interior of the Earth, where they found the horse at the hermitage of Kapila Rishi they were not nice about it to the rishi. Other Puranas offer a bit more detail. They tell that the Saga ras came upon a northern ocean, which they passed over, and that they then entered into the bowels of the Earth.

There are traditional Tibetan Buddhist beliefs regarding the city of Shambhala and the kingdom of Agartha, in which the city is situated. Specifically, some conversations held by Nicholas Roerich, a patron of culture, with various lamas and Tibetans as he travelled in that region with his wife in the 1920s have been recorded in various books beginning with those by Roerich himself, including 'Altai— Himalaya' (1929) and 'Shambhala' (1930).

Roerich wrote: "T remembered how, during our crossing of the Karakorum Pass, my scholarly guide Ladaki, asked me: "Do you know why there is such a peculiar upland up here? Do you know that in the subterranean caves here, many treasures are hidden, and that in them lives a wonderful tribe which abhors the sins of the Earth?" And again when we approached Khotan the hooves of our horses sounded hollow, as though we rode above caves or hollows. Our caravan people called our attention to this. When we saw entrances of caves, our caravaneers told us: "Long ago, people lived there; now they have gone inside; they have found a subterranean passage to that subterranean kingdom."

Here are some very important passages of a conversation which Roerich had with a Tibetan lama 5 n 1928.

Roerich: Lama, tell me of Shambhala.

Lama: But you Westerners know nothing about Shambhala.

Probably you ask out of curiosity; and you pronounce this sacred word in vain.

After some cajoling by Roerich, the Lama studied him and replied:

Lama: Great Shambhala is far beyond the ocean. It is the mighty heavenly domain. How and why do you people take interest in it? Only in some places, in the Far North, can you discern the resplendent rays of Shambhala. The secrets of Shambhala are well guarded.

Roerich: Lama, we know the greatness of Shambhala. We know the reality of this indescribable realm. But we also know about the reality of the earthly Shambhala...

We know some high lamas have visited Shambhala. We know the stories of the Bury at Lama, of how he was accompanied through a very narrow secret passage... So do not tell me about the heavenly Shambhala only, but also about the one on Earth. • Because I know that a real one exists on Earth. Lama, how does it happen that Shambhala on Earth is still undiscovered by travelers? On maps you may see many routes of expeditions. It appears that all heights are already marked and all valleys and rivers explored.

Lama: But as yet these people have not found all things so, let a man try to reach Shambhala without a call! You have heard about the poisonous streams which encircle the uplands. Perhaps you have even seen people dying from these gases when they come near them. Many people try to reach Shambhala, uncalled. Some of them have disappeared forever. ONLY A FEW OF THEM REACH THE HOLY PLACE, AND ONLY IF THEIR KARMA IS READY.

In 1998, Jan Lamprecht wrote about this subject matter in his book, 'Hollow Planets'. A Tibetan lama, who is a renowned teacher of Vajrayana Buddhism and a Tibetan doctor, lectured in the San Jose, California, area and made a reference to Agharta. His title is 'His Holiness Orgyen Kusum

Lingpap so it seems that he belongs to a certain lineage and might be privy to ancient information on the matter.

Lamprecht wrote: "While lecturing in San Jose, this lama stated that Agar that could be reached from India by flying northwards for 7 days. I would assume the lama 's references were to the speed at which the bird might fly. If that is so, then the average bird flying northwards from India for 7 days would easily reach the Arctic.   "Roerich's lama had stated that Shambhala lies in the Far North. Could this be a reference to the Arctic Ocean, as I had indicated in my previous book?! YOU CAN ALSO CLEARLY SEE A MAP OF THE INNER WORLD on the rear cover of my book entitled "GRAVITY, MATTER & SPACE TRAVEL" published in 2006 by 'Trafford Publishing', by purchasing it for

$ 50.00 plus shipping + handling, by calling them @ 1— 888—232—4444.

Reference was previously made to the book by Nicholas   Roerich, entitled "ALTAI—HIMALAYA", a travel diary, published by

'Adventures Unlimited'

Press 2001, first published in 1929, for sale @ $ 18.95 + s&h Upcoming is a brief review of this travel diary 'ALTAI—HIMALAYA'. Russian—born mystic, artist, philosopher, scientist, author and explorer Nicholas Roerich (1874—1947) was a real 'New   Renaissance' man just like this author is who is presently writing this review for you! To him and to me, beauty and "GOD'S TRUTHS" were the fundamental real truths. Roerich devoted his life to expressing and appreciating this primary aesthetic. In 1920 he travelled to the USA with an exhibition of his paintings and stayed on to continue his work of uniting the arts + sciences and establishing cultural and arts centers as he had done in his homeland of Russia.

In late 1923, Roerich set out on what would become a 5— year journey through some of the most beautiful and unforgiving landscapes in the world. The expedition travel led from Ceylon to India and then Kashmir; over the Himalaya and across Sinkiang; up to the Altai Mountains and on to Omsk; then to Irkutsk near Lake Baikal, down through Mongolia and the Gobi Desert to Tibet and back over the Himalaya.

Roerich's diary of the expedition, 'ALTAI—HIMALAYA, was published in 1929, and this reprint of the original includes 20 reproductions from the inspirational artworks he painted in his travels. The apt introduction is by his US contemporary, the sacred geometrician/ architect Claude Bragdon.

Roerich wrote his diary on horseback and in caravans on his journey so it has a freshness and immediacy, bringing to life the diverse peoples, places and experiences. The expedition faced many hardships, just like this author faced in his past experiences as a scientist. They were detained by Tibetan authorities for five months at 15,000 feet in summer tents amidst extreme cold and deprivation.

This is a great mystical travelogue, and Roerich's perspectives on sacred knowledge, very ancient traditions such as Shambhala and all of the cultural changes are very insightful and invaluable. Finally, I 'd like to restate that
'AGARTHA' is definitely located in the

'underground vicinity' of "SHIGATZE". UNDERNEATH
SHTGATZE LIES THE HIDDEN CITY OF SHAMBHALA. THE EXACT ENTRANCE HAS BEEN
BLOCKED BY LAMAS, only known to very few lamas that reside in Agar that in Tibet!

THE DEEP DWELLERS

RE-INVENTION, PARANOIA AND DECEPTION THE MASKS
MAY CHANGE, BUT THE PLAY'S THE SAME

The final 3 decades of the 20th century have been filled with rumors: conspiracy theories about alien abductions, plots for world domination shared between aliens and secret government, Illuminati cabals and the like. Several interesting variations on age— old theories have permeated the field of ufology and conspiracy literature alike, and in many places the two have overlapped and in fact have completely blurred into one. These latter areas are of particular interest and significance to the study at hand.

One of the most commonly reported manifestations of "alien invaders" are the EBE (extraterrestrial biological entities) type, commonly called "greys " These beings are the variety most often reported in abduction accounts, and they are highly ectomorphic with scrawny, underdeveloped limbs and bodies, oversized heads and extraordinarily large black eyes (some abductees have reported these to be actually "reptilian" or "birdlike" eyes with slit pupils, the "blackness" only a protective artificial film, like sunglasses) . They are from 3 to 5 feet in height, averaging around 4 feet tall.

Again, their very physiology gives away their origins r for large protruding eyes with large slit pupils and needing an artificial protective covering would be hard to equate with a race that has mastered, and perhaps been genetically prepared for, interplanetary or inter— stellar travel. Outer space is an extremely bright, radiation— filled environment. The type of eyes described are those of a creature that spends most of its time in the darkness, as they are designed for optimum light reception. The shaded coverings, for venturing out into the surface world, are really self—explanatory.   Similarly, their bodies, small and easily maneuverable through tight spaces, with small   surface area and a minimum of body—weight, are ideal for an underground environment. Their method of locomotion, generally described as " shuffling", with hips moving strangely or "sideways", is another indication that they have developed in a relatively cramped place or even one where tunnels are commonplace. They are often described as smelling "musty", "like a snake" or "like rotten eggs "

This type of entity allegedly abducts unfortunate human beings and conducts medical or genetic testing on them, sometimes removing sperm and ova for use in "hybridisation experiments" or for purposes unknown. The abductions almost always take place under cover of dark— ness (doubly helpful to the molesters, since the "harmful" sun has set and people are drowsy or asleep).

The victim is taken into a UFO (usually disc—shaped or shaped like a child's spinning— top) for experimentation and also for a form of indoctrination consisting of intense three—dimensional audiovisual presentations. The primary message of these presentations seems to deal with a concern over human destruction of the Earth's biosphere and biodiversity. This becomes even more telling when it is recalled that the first underground atomic tests were concurrent with the resurgence of "witnessed activity" of UFOs on a previously unprecedented scale. Those tests may have rattled more than a few windows in the desert!

During the medical aspect of the abduction, local anesthetics are seldom if ever used, and the victim is left with a type of post— traumatic stress syndrome which at first is marked by amnesia and mental distress or unease, sometimes followed by nightmares, self— imposed social isolation and eventually a total recall of the experience, either through natural recovery of memory or through hypnotic regression therapy.

Other common themes emerge as well. The entities are described by their victims as "drone—like", "robotic", clinical and so forth, and are also described as "reptilian" "lizard—like" or as having reptilian, birdlike or amphibian—like skin texture.

In many instances, the abduction experience moves to a "cavern city", cavern world" or " underground base " It is in these latter sub— terranean places that "hybrid beings having alien, human or other animal characteristics — have been reported as being seen in development, along with cannibalism and the torture of human beings. One such site is beneath Archuleta Mesa near Du Ice, New Mexico. In these pro— longed episodes, other beings enter the scene such as " hybrid " children who are frail, pale and ectomorphic but generally human in appearance. Some abductees have even said that these children look like fairies"!

Other entities are not so harmless, such as tall "human" type with an aristocratic "Aryan" look; these are generally referred to as "Nordics " in the UFO literature, and bear more than a slight resemblance to the "light elves" and Tuatha de Danaan of old. Are these the "fairy children", hybrids and "changelings", all grown up and hard at work?

Another type of entity is more sinister in both appearance and attitude, described as a "reptilian humanoid", "lizard man" or "reptoid" and ranging from 5 to 8 or 9 feet in height. These "reptoids" are usually characterized as being t' in charge" of the other types, but upon occasion are said to report to taller, even stranger entities that resemble skeletally thin, giant "greys" or even giant "mantids'' •Normal surface humans, paramilitary in nature or appearance, are also seen in these underground areas.

Another interesting factor is that all of the entities described go to great lengths to convince their captives that they have come from "far away", from distant stars, planets and "vibrational frequencies" or dimensions. They have come all this distance because they are "worried about humanity", but they put out their heartfelt propaganda as they stick a huge needle in a woman's abdomen or up someone's nose without even a local anaesthetic.

Yet the biology of all the different types or castes of abductors, as horrific as it might appear to superstitious human eyes, is essentially that of animal forms which are natural to the Earth: mammalian and reptilian. Obviously, it is very important to these beings that such a logical connection not be made. If there is even a shred of truth to UFO abduction accounts, then it is more than apparent that the abductors want their victims and humanity at large to believe that they are from "somewhere else". While humanity looks continually upward at enigmas in the sky, what is transpiring beneath our very feet? Additional folklore and literature parallels are apparent in the accounts described. The "grey s" are identical to the order— following, human—abducting, drone—like Gala tur and Ushabtiu of the Sumerian and Egyptian underworld mythologies. In Shetland Island folklore, " little men" who abducted people were sometimes referred to as t' grey neighbors" and the "greys" also bear a strong resemblance in head and torso structure to the crypt id El Chupacabras.

The methodical imps and Dj inn that served Satan and
Shaitan come to mind as well, and of course, Richard Shaver's "dero" are similar, especially in their use of high technology. The Nordics, sometimes seemingly the "enemies" of the grey and reptoid types, are often reported as working side by side with them in the underground facilities or labs, which recalls the fact that the Nagas were said to look "almost human" and reminds us of the mercurial dispositions of the aristocratic or Aryan faeries, the light elves/Tuatha de Danaan.

Additionally, the apocryphal hybrid offspring of the
Nephi lim and humans were described in the Book of Lamech and the Slavonic Book of Enoch as having an extremely Nordic or Aryan appearance, and were also said to have a reptilian patch of skin ("badge of priesthood") on their chests or elsewhere.

In Celtic Welsh myth, the lord of the underworld of Annwn had a magical cauldron that produced an endless supply of warriors for him. Was this an "Earth mother", primal womb archetype, or was it instead synonymous with producing offspring through gene—splicing, the cauldron" actually a test tube?

The reptoids or lizard men are familiar as Nagase Utukku,
Ammut, "dragon kings", goblins, trolls and so on. They are also often de— scribed as smelling like rotten eggs or like sulphur.

Both the reptilians and the grey s have in recent years passed from the UFO research realm into "conspiracy" literature, where they are said to be involved in "controlling the Earth" or are in the process of "taking over".

Several authors have published the theory that the reptilians are masters of illusion and holographic projection or are physical shape— shifters who are replacing world leaders, government officials and public figures as an insidious "fifth column". Here again are the changelings of European folklore, the hologram— utilizing dero and the "serpent men", precisely as described by the fiction writer Robert E. Howard in his pulp—fiction tales of King Kull. This is by no means a "new idea", revelation or suspicion, but is as ancient as the concept of an underworld itself. MEN IN BLACK

This short treatise would not be complete without an examination of other mysterious players from the depths: "Men in Black", or ' MI B'. These cool cats figure prominently in both UFO and conspiracy literature. Thanks to the excellent work of researchers and above all the late writers John Keel and the late Jim Keith, many small and seemingly inconsequential facts have been obtained from witnesses and preserved.

Men in Black seem to fall into 2 categories: the standard men in black, who may be actual agents of secret government or military investigation disinformation groups; and the MI B, who Keel indicates are somehow in league with or originating from the 'unseen interdimensional and various crypt i d' or creature sightings as well as the UFO phenomenon.

It is the latter category that has bearing here.

These "mystery men" are usually described as of varying height and build, mostly on the thin side. They usually wear dark or black businesslike clothing, dark hats and sunglasses
(again, the eye protection from the sun!). They're either olive—skinned and vaguely Oriental or Asiatic, or they' re Nordic or Scandinavian in complexion, hair color and physique. From the basic template of these 2 forms, strange variations have been reported: total hairlessness (not even having eyebrows or eyelashes) , overly large, protruding eyes (a non— mammalian trait, for the most part, or perhaps due to an unaccustomed lack of atmospheric pressure); wheezing and other trouble breathing, as if unused to earthly or surface air pressure? ; unnatural joint movement and locomotion; 'reptilian" or "froglike" cast of skin texture and facial features; webbed fingers; sulphurous or "metallic" odour; and a host of additional oddities of physical configuration.

Add to this the fact that these MI Bs are often ignorant or in amazement of the most ordinary surface—world activities trying to drink gelatin, refusing food and taking a pill instead, stealing or asking for small common objects (like writing pens) as apparently prized souvenirs — and they become even less human through their behaviour.

They often exhibit a strong interest in the sexuality or sexual habits of those they confront; and the longer they remain on the surface,

the more erratic, disoriented, lethargic or "drunk" they seem to be— come a side effect of rapid depressurization or aeroembolism.

They have sometimes identified themselves to their unwilling visitants with cryptic statements, e.g., claiming tozéitizens of the 'Nation of the Third Eye' an occult or secret society reference still widely used by 'reptilians', by some 'illuminati brotherhoods' of these enlightened human beings, but which harks back to the -' 3rd eye' of the Nagas or 'lizard' people, or the 'skull horns' of the Chinese dragons. BAPHOMET has a direct link to this Nation of the Third Eye, as a very definite parallel, having been in existence among 'shapeshifters' Another clue lies in the fact that they arrive at their victims' door— steps in shiny black cars that are in pristine condition, i.e., like new'. The puzzling thing is that these vehicles are almost always decades out of date and sometimes seem to be composites of several different makes of cars still out of date. To remain unravaged by the passage of time SHAPESHIFTING must be involved here, dear folks! Add to this the manner in which both the cars and the men suddenly and inexplicably vanish, as if swallowed up by the Earth. but by now the premise of 'SHAPESHIFTING BAPHOMETS' is very obvious. Many of the caverns and tunnels are damp worldwide, but many others are very dry and remain constant in temperature, year round, after a certain depth.

History and folklore both have parallels. Folklore is filled with 'dark men', "men dressed in black" and 'grim reapers', often identified in previous centuries and now as sorcerers, demons, warlocks or other servants of the devils + 'BAPHOMETS'. . .

During the plague years of the Middle Ages, entities resembling both MI B and the 'grey aliens' were often seen in areas that would shortly thereafter be stricken with an outbreak of the plague. On the eve of major events throughout history, people have repeatedly seen or been harassed by such beings.

As the late John Keel points out in the 'Disneyland of the
Gods' and his other excellent books, Julius Caesar, Napoleon and even Malcolm X all reported encounters with this variety of terrifying beings. Adolf Hitler also was alleged to have had his share of midnight visits from a mysterious 'Tibetan, and through him to have met "the New Man" — a sort of super—Aryan who he believed came from the interior of the Earth, and of whom he was most afraid. How much fear, confusion and human suffering can be traced to uncanny visits from these robotic, shape— shifting 'agent's provocateurs'? As I had mentioned previously, what is their long—term agenda? They do create the problems, are in charge of their goals to generate confusion and divisions worldwide. The shape—shifting 'Baphomet' invaders have been here for a very long time, along with the 'Jesuit' criminals that were in charge of the assassinations of President Abraham Lincoln, President John F. Kennedy, Robert Kennedy,
Martin Luther King and the son of President Kennedy 'John—John', just to name a few well—known public figures from past history 7 including Princess Diana....

THEY'RE DEFINITELY NOT EXTRATERRESTRIALS BUT
INTRATERRESTRTALS LIVING HERE ON THIS PLANET AND
INSIDE THIS VAST, UNKNOWN WORLD BENEATH OUR
FEET, stretching down through many very secret, twisting tunnels and deep caverns to the Mohorovicic layer, which itself is an anomalous cavern region deep beneath the upper crust which is about 30 miles in depth worldwide.

What if there is an unknown world beneath our feet   a world that is dependent upon the biodiversity and genetic wealth on the Earth's surface; a world that has been exploiting that wealth for thousands and millions of years, victimizing the ignorant savages who roam the face of the sunlit world? Or could all of the evidence be circumstantial and without merit, simply a misinterpreted conglomeration of coincidence and misidentified animals, natural phenomena and arche— types from -the human collective unconsciousness? The critic could toss in an endless supply of overactive imaginations down through the millennia, but the evidence dating back as it does for thousands of years of human traditions + continuing until our present days now says otherwise.

The reptilian, vampiric, robotic and demonic are all characteristics which have been attributed to underworld 'shapeshifting' beings and tricksters down through the ages. These beings are natural and technologically advanced to a degree that only until recently has been beyond our comprehension + considered 'magical Now we would do very well to become more aware not only of our own haunted planet but also of our mythic and folkloric heritage, for it speaks not of a symbiosis but of many 'phase— shifting' entities co—existing with the numerous and 'invisible parasites' dwelling in the depths and on the surface areas of "THIS HAUNTED" PLANET EARTH, AS WE HAVE SEEN TN THIS BOOK!

DID JAPAN'S NIGHTMARE DISTORT THE FABRIC OF SPACE-?
TIME?

NUCLEAR MELTDOWN & TIME DILATION

This is a transcription of an article that was written by C, C, von Werklaåg, a science writer. The readers will find this really unique!

If, at this moment, you were to contact someone in Japan to inquire "What time is it now? ", you could reasonably expect a response of exactly 13 hours' difference (if you live in the U.S. Eastern time zone). If it's 3.30 p.m. in New York City, then it's 4.30 a.m. the next day in Tokyo. From the perspective of those of us living in North and South America, it appears that the Japanese live in the future, but how much more surprised would you be to learn that, not only is the hour different, but the era is also inconsistent with your perceived present? Instead of hearing someone say "It's four—thirty you would be shocked to learn that it is also "1932."

Such scenarios exist throughout fantasy literature and films, yet many contemporary researchers agree that what was once only within the realm of science fiction is quickly becoming science fact. The prominent Tokyo physicist, Risa Imai, claims that many Japanese citizen; have experienced what she calls "dilatory ripples" in the fabric of the time—space continuum. Professor Imai, as well as several other scholars around the world, believe this phenomenon results from the massive meltdown of the nuclear reactors at the Fukushima Daiichi facility, a process which began with the devastating earthquake and tsunami that rocked Japan on 10 March 2011.

Dilations in time occur when independent observers simultaneously perceive noticeable differences in objective reality. Such observations are generally due to some alteration in space— time resulting from differences in the rate of travel of independent observers, such as that which occurs during the famous thought experiment called the '1 Twin Paradox."

In this instance, one identical twin leaves the earth in a space— craft traveling near the speed of light while his brother stays behind. The astronaut travels throughout the galaxy for several years, return in to find that while he has aged only a few light years, his twin is now an old man.

Although some scientists speculate that the sort of time travel Davies mentions requires the warping of space— time via a black hole, others, like Professor Imai, feel that such radical changes in the flow of magnetic energy could be disrupted enough to offer brief door— ways through which one might peer or perhaps even slip through for brief periods, Imai notes the tremendous quantities of radiation that leaked from the Fukushima power plant facility in the weeks prior to its total collapse and how the introduction of "uncontrolled radio— active stimulation" results in "a destabilization of geophysical conditions, including gravity" (page 14 of her report to the Japanese political group, the Diet). A Reuters article for April 12, 2011, describes the leak at the Miyagi prefecture facility as equivalent to

"a tenth of the amount of radiation released in the

Chernobyl disaster" (para 1). A confidential source at
Tokyo Electric Power confided that the "reported

amounts of radiation leaked during the course of the facilities' meltdowns was "conservative in the extreme.

While human beings are constantly exposed to various forms of radiation the effects are often negated by the planet's atmospheric shielding. But some earthbound phenomena do create situations where exposure may result in radical effects just as deleterious to life as any nuclear disaster. Shannon Pal discusses how small— scale gamma ray bursts occur regularly on earth via "terrestrial gamma—ray flashes" associated with lightning strikes and notes thunder and lightning can emit far more energy than previously thought and release streams of antimatter particles" (Discover, April 2011). Such power might lend itself to a disruption of matter down to the level of quanta or introduce enough analogous energy into the earth's own natural system to make quantum teleportation possible (see Rules for a Complex Quantum World by Michael A. Nielsen, Scientific American November 2002).

Professor Imai 's investigation revealed several cases in point, including a group of survivors of the earthquake and the tsunami whose small coastal city of Sendai was flattened. When they were eventually allowed back into the area in order to check on the devastation and to search for lost family members, an entire van load of people reportedly experienced an unusual phenomenon. Each stated that she/he had driven through an oddly shaped "kumo" or cloud. The object appeared faintly luminescent (even though it was the middle of the day) and seemed to elongate around the van as the vehicle moved through it. For several moments (one witness swears he measured the duration at just over 4 minutes), everyone in the vehicle, including the driver, could see what looked like an ancient village on either side of them. Toiji Yakamura, a Civil Defense engineer who accompanied the party into the disaster area, described what he saw as "Miyagi prefecture, only as no one has seen it for quite some time." Mr. Yakamura said, "We 've all seen photo— graphs from the early Victorian period with men dressed in stovepipe hats walking arm— in— arm with women in kimonos and face paint. These styles co—existed with the more traditional fashions worn by the samura: who were still around back then. It was this which we all saw. We all thought we'd accidentally driven onto a movie set." But no film crews were working in the area other than those reporting on the crisis at the nuclear plant. Mr. Yakamura and his companions claimed that the images wavered, yet they all saw a village with buildings still intact and with people moving about, alive. The government agent driving the van (who remains nameless for security purposes) states that the cloud eventually caught up" with the vehicle, that there was loud kind of thump, as if we'd run over debris large enough to rock the springs" and then the scenes around them changed. One minute they were looking out on what he called "that ghost world," while the next there was little more surrounding them than ruined fields and streets filled with the shattered remains of what had once been an enchanting fishing community Professor Imai cites several other instances where individuals encountered portals into alternate histories and having glimpsed scenes from feudal times up through what they thought looked like the 1940s. Most people claimed to see the images for brief moments without experiencing the kind of interaction reported by the group from

Sendai. However, Mrs. Junko Fuj iwara was terror stricken when, on March 15 her husband, Hisekazu, stepped through a similar window as others have described and was swallowed up. After only a few seconds, the opening slammed shut with what she described as a "clap of war a says that her husband has yet to return. Risa sidebar to the story — a conversation with Tokyo who recalled another case involving a man with the thunder. Mrs. Fuj i — Imai adds a peculiar police detective same name. The detective says that an individual calling himself Hisekazu Fuj iwara stumbled into the police station insisting that he had walked through an opening in time to the 1960s and then another into the "present " era. The detective remembered the exact date, March 15, 1981, because it was his first official day as a rookie police officer. Could it have been the same person? If so, he missed the proper time frame by 30 years, or perhaps a version of him from another dimension fell through to this side.

Physicist Andre Linde's award—winning work on zero— point dark energy presumes that our universe is inexorably linked with a mirrored twin, one which is diametrically opposite in its energy values. This close proximity to alternative realities fuels conjecture for mathematicians like Brian Leno and Ward Locke, who concur with Professor Risa Imai 's research in believing that the areas where one universe (or at least one level of its dimensional presence) converge at points where space— time has worn thin. Leno and Locke's calculations lead them to assume that such "melds," or portals, cannot be sustained in perpetuity due to gravitational collapse, which they claim is a byproduct of the instability of "equalinear regionality" or in layman's terms, two realities coexisting in the same space; this state provides insufficient stability to allow for transference between dimensions, meaning that it is impossible for individuals to do more than "see into" the other state not "step into it whole". However, Professor Imai 's calculation indicate that the introduction of antimatter into the equation, such as one might find in the phenomena described by Shannon Pal us, boosts the dynamic range of the respective gravitational states to levels which permit stability in localized "subverses" (Imai rationalizes these as microcosmically compartmentalized areas of our own and/or other universes). These momentary windows might allow a person, or smart group as she described earlier, to cross over from one plane of reality to another.

Professor Imai is not alone in recording time dilation anomalies of this nature. In 1986, shortly after the meltdown of the nuclear core at Chernobyl, Sergei Illyvich, working for the KGB's Atomic

Science Division (currently employed as consultant with OK B Gidropress for their nuclear facilities) produced several dossiers concerning phenomena similar to that described by Professor Imai 's research. Illyvich 's findings, only recently released through an equivalent of the U.S. Freedom of Information Act, revealed numerous instances where citizens witnessed temporal variants particularly scenes from Tsarist Russia and from the Napoleonic Wars and consequent loss of life (presumed) when people stepped through portals into these dilations. Illyvich 's reports also note a peculiar difference to Professor Imai 's findings instances where objects, not people, came through from antiquity to the present. The most striking of these occurred when a full— sized Zeppelin airship, presumed lost in WW1, floated into view and crashed on the outskirts of Pry pi at, not far from the Chernobyl facility. No crew was aboard, but like many other "ghost ships" described throughout history, the craft showed signs of recent occupancy with proper gear and clothing for the time period and the remains of a fresh meal reportedly strewn throughout the debris of what was once the officer's mess hall.

Interestingly, both Illyvich and Imai, working decades apart, concluded that the sudden and massive release of nuclear radiation brought about the dilatory effects in the temporal reality of the regions findings substantiated by credible outside observers living near or traveling through the affected areas.

Of course, if what Michael Talbot concludes is true, then the joke may be on all of us. In 'Beyond the Question', Talbot proposes the concept that "without observers it must be accepted that the universe would not exist. " This challenging implication points out how, in the past decade quite a number of amazing coincidences in the law of physics, coincidences that imply the universe was designed for the purpose of creating conscious entities capable of observing and understanding it, have come under scientific scrutiny, and currently there is an active debate about what these amazing coincidences mean. Some scientists believe that the human race, through billions of acts of observerparticipancy traveling back through time, has actually had a major role in creating both the universe and the laws of physics.

Thus, one might conjecture that neither a convergence of dimensions or introduction of radioactively induced gravitational changes are necessary for temporal discrepancies. Through powers derived from our own thoughts and observations, we might simply be doing all of this to ourselves.

ANCIENT TUNNEL FOUND IN TEOTTHUACAN

In Mexico's ancient city of Teotihuacan archaeologists have discovered a mysterious tunnel underneath the Temple of the Serpent. Running to the east, the tunnel is about 130 yards long.

Sealed off for more than 1,800 years, the passage is 45 feet beneath the surface and decorated with symbols of the underworld. Researcher Sergio Gomez Chavez with Mexico's National Institute of Anthropology and History reported the discovery originally detected with ground penetrating radar. The tunnel leads, it is believed, to chambers pro— viding the last resting place of ancient rulers. Gomez Chavez says the discovery could be one of the most important of the 21st century.

WEB SURFER DISCOVERS "SECRET BASE ON MARS" • • • OR NOT?

One internet—surfing interplanetary explorer has come up with a discovery which for a few days in June set the worldwide web on fire with talk about a secret Martian base.

David Martines posted a You—Tube video of a Google Mars flyby and zoom—in which showed a large white anomalous object on the Martian surface. The white tube—like structure he called "Bio—Station Alpha, " and he soon had hundreds of thousands of downloads, and mentions in the tabloid press. Coordinates for the location are 49 '19.73"N 29 33 '06 53"W.

Later he says, he learned that the feature was some kind of ice formation and he posted a new video to YouTube, which hasn't been nearly as popular as the first.

The ice itself, though, is unusual, and must be included among the many anomalies which have been sighted on the Martian surface, and which have yet to be satisfactorily explained.

THE TRUTH ABOUT MARS is a LANDMARK PIONEERING

BOOK ABOUT LIFE ON MARS

"THE TRUTH ABOUT MARS" answers the question: Is there human life on other terrestrial planets in our galaxy? It also addresses the nature of man's consciousness and man's ability to make use of the electronic design of the mind in the exploration of the universe.

This book is not about the paranormal nature of life; it is the statement of one man's experience, validated by thousands of photographs returned by the Mariner, Viking, Pathfinder, and Global Surveyor space probes.

Yet the basic inquiry requires man's recognition that he carries within himself the substance which is recorded in so—called space and time. When we can assimilate this, it will serve to remove the amnesia that has blocked our memory of the reality of extraterrestrial life on other than one earth planet, in one solar system, in one small galaxy, in one universe.

The late Ernest L. Norman deals with an astral flight through space. His Martian guide Nur—E1 takes him on a tour of the underground cities of Mars, Martian Science and Technology, the Martian Society, the Harmony of Martian Life. The Supernova Connection deals with destruction of surface life prior to the Martians moving underground.
Martian Scientists developed artificial wombs in their labs on Mars. It deals with new evidence of prior habitation on the unearthly land— scape on Mars and the global climatic changes on Mars. There is much evidence for liquid water on Mars with its polar icecaps constantly changing during the year The surface life on Mars went from deluge to desert due to the supernova disaster a very long time ago.

Appendices deal with 'Mars Pathfinder' Results from
JPL.

Many pictures boost the view that Mars supported life a very long time ago! New Global Surveyor Data reveals deeply layered terrain, magnetic features, and genesis of a Martian dust storm. Enclosed are photos dealing with Mars. It includes view of a street in 1 of the Martian underground cities, which are inside the metal tubes that have been constructed one mile below the surface of the planet, and other views!

For the coming years through 2042, we will begin to see, in effect, a wider window into the universe and, an opening into the "cosmic curtain" that heretofore hid the people of planet Earth from the real it* of other terrestrial civilizations in our galaxy.

When additional information from future probes launched by NASA is examined, we will realize that planet Earth is not the sole proprietor of Homo Sapiens.

Throughout our history, and from records of the civilization of Atlantis, there have been records of scientific research, left by individuals in different disciplines of knowledge, which have formed a tapestry depicting the prehistory of humankind on Earth. This history has been woven into the mythos traceable to varied cultures and societies throughout our planet.

History reveals that mankind has always been searching for the under— standing of life, indeed, yearning for knowledge of the cosmic scheme and our position in the design called Life. Planet Earth is, in reality, a research laboratory for the billions of people who inhabit its surface and the hundreds of millions of other specimens' plants, animals, insects, birds, and the watery ecosystem that provides the environmental basis for survival on Earth.

We humans, curious about the unknown which lies beyond us, have looked outwardly from our small planet, longing for knowledge of other planets in the immensity of space. It may be because of our inquiring mind that we are always seeking for the logic and reason for our existence in an immense universe and for our purpose in the Cosmos!

The desire for a logic and reason for the existence of life validates that the human mind is, itself, constructed from the stuff of the universe, the energy that has fashioned the planets, stars, galaxies, universes and humankind.

The cosmic curtain is indeed opening as we come to the beginning of a new millennium in 2012, which itself is a sub-cycle of the 6th cycle of the recessional. 2012 and the future years to come will determine our future at many different levels! After all, life is for learning, and learning is indeed one's life—long purpose, to advance the logic and reason for one's involvement in life and to further the purpose of progressive evolution and our attunement with the Superconscious within our GODGIVEN RIGHTS ON THIS PLANET EARTH. GOD BLESS YOU ALL!

ADVANCED LIFE ON MARS AND VENUS

Cosmic visionary Ruth Norman of the Unarius
Foundation informs us in her book 'MARS
UNDERGROUND CITIES DISCOVERED' that Mars was the victim of a huge disturbance some 160,000 years ago, described as a Nova but involving the entry into our solar system of a body passing close to Mars, colliding with Earth, and destroying the continent Lemur i a. We shall return to this but in the meantime let us take a look at contemporary sciences' information on Mars.

Scientists continue their age—long quest for the answer to the riddle of whether there is life on Mars. In 1965 Mariner TV reached the vicinity of Mars' surface and transmitted pictures back to

Earth. Fuzzy pictures revealed a barren planet with lunar—like craters indicating a surface unchanged in a millennium.

In 1969 two spacecraft were sent to Mars revealing still more austere and hostile conditions and the unlikelihood of life. Undaunted and far from satisfied, scientists launched Mariner TX which went into orbit around Mars in 1971, but on the side of Mars not previously seen 4 huge volcanoes were apparent which dwarfed our highest mountains. Furthermore, a "Grand Canyon" was revealed some 3000 miles long and 15 20,000 feet deep.

Many geologists are convinced that the 30 mile—wide canals first seen by Mariner IX were formed by giant floods caused either by meteors impacting and melting ice under the surface of Mars, or volcanic activity. However, an immense amount of water movement is required to explain the massive rifts and plains.

Now, the so—called planetary "Nova" which destroyed Lemur it affected Mars by dehydrating and devastating its surface. Nevertheless, the civilization even then was sufficiently advanced and in communication with the spiritual hierarchy to receive prior information of the intruding body and prepare by building cities underground.

The nova was actually a manifestation in the third dimension resulting from higher— dimensional energies of a positive kind directed at the negativity enshrouding planet Earth in order to cancel it. The positive and negative energy interaction expressed itself in the apparent event of a so— called planetary body entering the Solar System. Thus the Mars' civilization continued, and according in Norman's book the civilization is still in existence, underground. Taking into account other data, the indications though are that it is no longer in our dimensions. This question is not made clear, but this is supported by the fact that out— of—body visits (overshadowed by the spiritual hierarchy) have been made by Unarius channels from whence came much of the information, and also other sources such as the fact that the spiritual hierarchy directed the massive negative thought form remaining after the destruction of Maldek — a civilization on a planet beyond Mars of which the fragments now are the scientifically unexplained asteroid belt between Mars and Jupiter and attached it to Mars to keep it away from Earth.

While on this interesting point another reason for Mars being associated with war is that prior to the destruction of Lemur i a, Lemurians, using their space fleet, launched an all—out attack on Mars. The inhabitants were prepared and using a surprise strategy defeated the Earthmen. However, propaganda continued to the effect that, in fact, it was Mars which attacked Lemur i a, thus giving Mars the reputation of being very hostile.

In our context then it appears that this Mars civilization is no longer in our physical dimension. This would also tend to be supported by the fact that this civilization has continued for hundreds of thousands of years without destruction, implying great spiritual development has been taking place. As the frequency of consciousness rises so the environment eventually passes beyond physical perception

Nevertheless, the existence of the Martian civilization is still quite physical and involving corresponding material needs of man. The "canals" of Mars are supposed to be the surface ground effects (or after effects) of the great interconnecting transport system tunneling from one city to another.

A typical city was about one mile in diameter consisting of beautiful buildings and sculptured gardens, fountains, fruit trees, statues, murals and, in general, a horticulturist' paradise. The enclosed roof area of these underground cities appears as a sky with simulated twinkling stars even a "moon" is revealed periodically during the night cycle, and for day—time, artificial sunlight is used. Transport is by shuttle craft or moving sidewalks. These people have developed their consciousness to a high level of sanity, intelligence and love for all creation. Their bodies are described as about an average of 5 feet, 6 inches and are similar in appearance to our Chinese race.

Advanced technology enables children to learn mainly by video with the addition of sleep learning in which a ray transfers knowledge subconsciously. Birth has an alternative process to the natural one. The fetus can be developed externally in special vessels if the mother wishes.

The Martians can travel between galaxies in their spacecrafts but even with their advanced technology enabling them to travel hyper spatially it takes them years. Such ships as theirs may contain as many as 5000 passengers, although they refer to encounters with city ships from other civilizations containing up to one million people.

Martians, it is explained, did visit Earth and set up a colony in the Gobi Desert.
However, the primitives of Earth were found to be so backward and hostile that this

Martian settlement was abandoned. Thus we see then that Mars has a very advanced civilization but seemingly not any longer in this dimension.

Planet Venus presents a similar situation but is even more spiritually advanced and definitely not in our dimension.

The surface of this planet has always caused a mystery to astronomers, with its perpetual veil of clouds. The information given in Ernest Norman's book 'The Voice of
Venus' is that this blanket of vapor is not a natural occlusion. The surface has been shielded
by Venusians deliberately by means of high—frequency energy, with aetheric conden— sati
on forming an envelope around the planet; the purpose being to prevent man's confusion in
advancing science's possible detection of inexplicable elements and strange radiations.

There are, in fact, magnificent civilizations on Venus which exist at higher vibrations of the atomic vortex and correspondingly greater dimensions. This places their existence outside the material 3D band of reactionary energies and thus not vulnerable to the severe physical or 3D environmental conditions on planet Venus. However, all levels and dimensions interact and there will be correspondences between the higher level atmosphere and the physical one. But even at the physical level on Venus there is consistency, good behavior of weather conditions and no snowy regions owing to the planet's axis not being tilted, as is Earth's, creating the seasons. These higher frequency civilizations on Venus enjoy an environment of iridescent glowing beauty and brilliance of color indescribable to earthlings: radiant opalescence of oceans, rich bounteous vegetation, crystal mountains and harmonious animal, bird and insect life. There are societies which are at different levels of the scale of consciousness and exist separately in different cities. In general, how— ever, inhabitants will encounter crystal cities, fairy—like castles, forest glades, environments lighted by the perpetual radiance of the Infinite, rainbow hues reflecting sparkling shafts of

radiant light, and an invigorating "atmosphere" free from pollutant energies.

Structures such as buildings or one's own home are created by materialization through psychokinetic abilities. This gives rise to great variations in design or architecture. The more advanced cities are on mountain tops, and the least developed in the valleys or on the flat plains. As we have mentioned, these lower levels of consciousness are

well above those of Earthlings, but nevertheless many of
these beings still have ties with materialistic past experiences and feel the need for corresponding experiences, such as eating.

Buildings may be enormous with extremely spacious interiors, although average height of the Venusian is not much greater than Earthman. Cities may present a beautiful array of domes, minarets and spires, thoroughfares adorned with plants, and blossoms befitting a horti— culturalist's dream.

The construction, positioning and shapes of certain centers are architecturally designed to filter and regulate influx of cosmic energies for meditative purposes and communication with other worlds or universes. In particular, Venus is noted for its dispensation and ad— ministration of healing energies and consequently is referred to as the mother planet for the Solar System and is the most spiritually advanced planet in our system.

Reference: This part was transcribed by this author from

"ESCAPE FROM THE UNIVERSE" by the author of   'SUPERHUMAN', and 'THE SECRETS OF FLYING SAUCER PROPULSION' NOEL HUNTLEY.
This book is now out of print, originally published in 1985.

www.ingramcontent.com/pod-product-compliance
Lightning Source LLC
Chambersburg PA
CBHW050713180526
45159CB00003B/1016